ALTITUDE TRAINING AND ATHLETIC PERFORMANCE

Randall L. Wilber, PhD, FACSM
United States Olympic Committee

Human Kinetics

Library of Congress Cataloging-in-Publication Data

Wilber, Randall L., 1954-
   Altitude training and athletic performance / Randall L. Wilber.
     p. cm.
   Includes bibliographical references and index.
     ISBN 0-7360-0157-3 (hard cover)
   1. Altitude, Influence of. 2. Physical education and training. I. Title.
   QP82.2.A4W54 2004
   613.7'11--dc21
                      2003014329

ISBN: 0-7360-0157-3

Copyright © 2004 by Human Kinetics Publishers, Inc.

All rights reserved. Except for use in a review, the reproduction or utilization of this work in any form or by any electronic, mechanical, or other means, now known or hereafter invented, including xerography, photocopying, and recording, and in any information storage and retrieval system, is forbidden without the written permission of the publisher.

The Web addresses cited in this text were current as of July 2003, unless otherwise noted.

Acquisitions Editor: Michael S. Bahrke, PhD; **Developmental Editors:** Rebecca Crist and Jennifer Clark; **Assistant Editor:** Derek Campbell; **Copyeditor:** Joyce Sexton; **Proofreader:** Jim Burns; **Indexer:** Marie Rizzo; **Permission Manager:** Dalene Reeder; **Graphic Designer:** Andrew Tietz; **Graphic Artist:** Kathleen Boudreau-Fuoss; **Photo Manager:** Kareema McLendon; **Cover Designer:** Jack W. Davis; **Photographer (cover):** Stuart Hamby; **Photographer (interior):** Courtesy of USOC, unless otherwise noted; **Art Manager:** Kelly Hendren; **Illustrator:** Craig Newsom; **Printer:** United Graphics

Printed in the United States of America   10 9 8 7 6 5 4 3 2 1

Human Kinetics
Web site: www.HumanKinetics.com

United States: Human Kinetics
P.O. Box 5076
Champaign, IL 61825-5076
800-747-4457
e-mail: humank@hkusa.com

Canada: Human Kinetics
475 Devonshire Road Unit 100
Windsor, ON N8Y 2L5
800-465-7301 (in Canada only)
e-mail: orders@hkcanada.com

Europe: Human Kinetics
107 Bradford Road
Stanningley
Leeds LS28 6AT, United Kingdom
+44 (0) 113 255 5665
e-mail: hk@hkeurope.com

Australia: Human Kinetics
57A Price Avenue, Lower Mitcham
South Australia 5062
08 8277 1555
e-mail: liaw@hkaustralia.com

New Zealand: Human Kinetics
Division of Sports Distributors NZ Ltd.
P.O. Box 300 226 Albany
North Shore City, Auckland
0064 9 448 1207
e-mail: blairc@hknewz.com

To the memory of my father,
Robert W. Wilber,
a man who will be remembered
for his intelligence, love of nature,
compassion, blue-collar work ethic,
integrity, and, above all,
the tremendous love and devotion
he had for his family . . .
a unique and special man
whose life was well-lived and well-loved.

CONTENTS

Preface   ix
Acknowledgments   xiii
Introduction: A Brief History of Altitude Training   xv

Part I — The Science of Altitude Training

Chapter 1 Potential Physiological Benefits of Altitude Training   3

Hematological — 3
Skeletal Muscle — 13
Genetics and Altitude Training — 14
Summary — 15
References — 17

Chapter 2 Physiological Responses and Limitations at Altitude   21

Arterial Oxyhemoglobin Saturation — 21
Maximal Oxygen Consumption — 24
Aerobic Performance — 33
Training Capacity — 42
Anaerobic Performance — 46
Cardiovascular Responses — 50
Ventilatory Responses — 52
Respiratory and Urinary Water Loss — 53
Blood Lactate Response and Acid-Base Balance — 53
Carbohydrate Utilization — 55
Iron Metabolism — 56
Stress Hormone Response: Cortisol — 57

Immune Function — 58
Oxidative Stress — 60
Sympathetic Nervous System: Epinephrine and Norepinephrine — 61
Skeletal Muscle Adenosine Triphosphatases — 63
Body Composition — 63
Female-Specific Physiological Responses at Altitude — 64
Sleep Disturbances — 65
Acute Mountain Sickness — 66
Sickle Cell Trait and Exercise at Altitude — 66
Cognitive Function — 67
Summary — 67
References — 68

Part II Altitude Training and Athletic Performance

Chapter 3 Performance at Sea Level Following Altitude Training — 83

Factors Affecting the Results of Altitude Training Studies — 83
Traditional Live High-Train High Altitude Training — 85
Contemporary Live High-Train Low Altitude Training — 105
Summary — 111
References — 115

Chapter 4 Performance at Altitude Following Acclimatization — 119

Early Research (1967-1975) — 119
Recent Research (Post-1990) — 122
Summary of the Scientific Research — 123
Practical Application — 124
Preparing for Competition at Altitude: The U.S. Olympic Team Experience at the 2002 Salt Lake City Winter Olympics — 127
Pre-Acclimatization — 133
References — 134

Part III Practical Application of Altitude Training

Chapter 5 Altitude Training Programs of Successful Coaches and Athletes ___ 139

Altitude Training Camps — 140
Training Programs Used by Permanent Altitude Residents — 159
Contemporary Altitude Programs That Utilize "Live High-Train Low" Techniques — 175
Summary — 180
References — 182

Chapter 6 Current Practices and Trends in Altitude Training ___ 183

Normobaric Hypoxia via Nitrogen Dilution — 183
Supplemental Oxygen — 193
Hypoxic Sleeping Units — 206
Intermittent Hypoxic Exposure — 207
Ethical Considerations — 217
Summary — 219
References — 219

Chapter 7 Recommendations and Guidelines ___ 225

The "Million-Dollar" Questions — 225
Training Modifications — 226
Away From Training — 228
Altitude Training Sites — 228
Altitude Training Resources — 228
References — 231

Index   233
About the Author   240

PREFACE

"Does altitude training lead to improvements in athletic performance?" That question has been asked by athletes, coaches, and sport scientists for many years. Data from several scientific studies have failed to conclusively answer the question. Anecdotal evidence, however, suggests that altitude training can enhance athletic performance. Altitude training and its effect on athletic performance continues to be a controversial issue among athletes, coaches, and sport scientists. Given the current controversy on this topic, the purpose of this book is to examine both the scientific and anecdotal evidence pertinent to altitude training and its effect on athletic performance. This book is not intended to provide *the* definitive answer regarding the efficacy of altitude training. Rather, it is designed to evaluate the efficacy of altitude training from the perspective of both scientific evidence and "real-life" practice by coaches and athletes.

The introduction to this book provides some historical perspective and background on altitude training. It focuses primarily on how the current interest in altitude training originated with the 1968 Mexico City Olympics, and introduces the question of why athletes from altitude-based countries like Kenya and Ethiopia tend to dominate endurance events in the sport of track and field.

Part I, "The Science of Altitude Training," includes two chapters devoted to the physiological aspects of altitude training. The purpose of part I is to provide the reader with an understanding of the physiological responses and adaptations of humans at altitude, as well as the ways in which some of these physiological adaptations may enhance athletic performance. Chapter 1 describes the scientific theory and rationale for using altitude training as a means of improving athletic performance. It is believed that aerobic performance may be enhanced following altitude training as a consequence of improvements in red blood cell mass and hemoglobin concentration. At present, there is inadequate scientific rationale to support the use of altitude training for the enhancement of most anaerobic events. Chapter 1 is intended to serve as an introduction for athletes and coaches and to provide a brief review for students and sport scientists. Chapter 2 describes the physiological responses and potential limitations that athletes experience when living and training at altitude. For example, athletes typically have to reduce their normal sea level training volume or training intensity, or both, upon ascent to higher elevations. This topic and others are addressed in chapter 2, which will be of primary value to athletes and coaches interested in including altitude training as a part of their overall training program.

Part II of this book, "Altitude Training and Athletic Performance," includes two chapters that address the question "Does altitude training lead to improvements in athletic performance?" Chapter 3 presents a comprehensive and detailed review of the scientific literature pertinent to the topic of altitude training and athletic performance at sea level. This chapter is organized into several sections focusing on scientific studies that have examined the effect of traditional "live high-train high" (LHTH) and contemporary "live high-train low" (LHTL) altitude training on sea level endurance performance, as well as their effect on hematological variables and skeletal muscle factors. Chapter 3 is intended to serve as a comprehensive review and reference for students and sport scientists, but will also provide athletes and coaches with a detailed base of scientific information on altitude training and its effect on sea level athletic performance. Chapter 4 summarizes both the scientific evidence and practical application pertinent to the topic of aerobic performance at altitude following altitude acclimatization. The information provided in this chapter is particularly timely given the 2002 Winter Olympics hosted by Salt Lake City, which is located at an altitude of 1,250 m (4,100 ft). Some of the events at the Salt Lake City Olympics were held in the Soldier Hollow area located in the mountains east of Salt Lake City at elevations ranging from 1,750 m to 1,800 m (5,740-5,900 ft).

Part III of this book, "Practical Application of Altitude Training," includes three chapters devoted to practical and applied methods of altitude training. Chapter 5 is based on the anecdotal evidence that supports the use of altitude training. It describes the altitude training programs of several coaches and athletes who have attained success in World Championship or Olympic competitions. The chapter summarizes what these successful coaches and athletes consider the optimal altitude at which to train, the optimal duration of an altitude training camp, the optimal training regimen at altitude, the time to return to sea level, and so on. Essentially, chapter 5 is intended to serve as a valuable "how-to" guide for athletes and coaches interested in including altitude training as a part of their overall training program. Chapter 6 describes several unique, state-of-the-art altitude training strategies and devices currently being used by elite athletes to improve performance. Most of these strategies utilize the LHTL model of altitude training and include normobaric hypoxia via nitrogen dilution (nitrogen apartments), supplemental oxygen training, hypoxic sleeping units, and intermittent hypoxic training. This chapter will be of value to sport scientists, athletes, and coaches interested in using the most current applications of altitude training. Lastly, chapter 7 provides a summary of recommendations and guidelines for altitude training based on the scientific and anecdotal evidence presented in this book. It also includes supplemental information such as a list of major altitude training locations throughout the world. References provided at the end of each chapter will be of primary value to students and sport scientists.

In summary, altitude training and its effect on human athletic performance is a controversial topic among athletes, coaches, and sport scientists. The purpose of this book is to consolidate the current body of knowledge, both scientific and anecdotal, relevant to the topic of altitude training and athletic performance. This book is unique in that it is written from both a scholarly and an applied perspective and thus should be of interest to sport scientists, students, athletes, and coaches alike. It is the author's hope that readers, regardless of their background or interests, find *Altitude Training and Athletic Performance* informative, intriguing, and enjoyable.

ACKNOWLEDGMENTS

I gratefully acknowledge Human Kinetics for their interest and support of *Altitude Training and Athletic Performance*. In particular, I wish to thank Jennifer Clark and Rebecca Crist, developmental editors, and Derek Campbell, assistant editor, for their advice, encouragement, and ceaseless energy in bringing this work to publication. I sincerely appreciate their professionalism and commitment to excellence. I also thank Joyce Sexton, copyeditor, Kareema McLendon, academic photo manager, Dalene Reeder, academic permissions manager, and Heather Culbertson, exhibits and marketing assistant, for all their hard work and valuable contributions to this book. Finally, I acknowledge Dr. Mike Bahrke, acquisitions editor, for recognizing the need for a book devoted to the intriguing topic of altitude training. He has served me well as both a professional colleague and friend.

I also wish to thank the athletes, coaches, and sport scientists for providing unparalleled expertise and insights on the various topics addressed in *Altitude Training and Athletic Performance*. This group represents some of the best in the world in their respective fields, and I commend them on their many years of athletic and academic excellence. It has been an honor and a pleasure working with them.

INTRODUCTION: A BRIEF HISTORY OF ALTITUDE TRAINING

This introduction provides some historical background and perspective on the use of altitude training for the enhancement of athletic performance. Most of the current interest in altitude training can be traced back to the 1968 Summer Olympic Games, which were held in Mexico City at an elevation of 2,300 m (7,544 ft). At the 1968 Olympics, sprinters and jumpers in the sport of track and field set several world records in the "thin air" of Mexico City, whereas the distance runners ran markedly slower compared with 1968 world records. In addition, athletes from altitude-based countries such as Kenya and Ethiopia won a relatively high percentage of medals in the middle and long distance races. Interest in altitude training has continued to grow since the 1968 Summer Olympics. A number of athletes, coaches, and sport scientists have attempted to determine how to use altitude training optimally for the purpose of enhancing performance.

Why Altitude Training?

Why do many elite athletes, particularly endurance athletes, use altitude training in preparation for important competitions? By living and training at altitude, athletes expect to get an increase in their red blood cell mass and hemoglobin. In turn, increases in red blood cell mass and hemoglobin have been shown to enhance an athlete's oxygen-carrying capacity and allow the athlete to train and perform more effectively upon return to lower elevations. This mechanism is described in more detail in chapter 1, "Potential Physiological Benefits of Altitude Training."

For athletes and coaches, the most important criterion for evaluating the effectiveness of a training program is performance. At the Olympic level, differences in performance are miniscule (Hopkins et al. 1999). In track and field events, for example, the difference in performance between Olympic medalists is typically less than 0.5%. Figure I.1 shows the finishing times of the medalists and fourth-place runner in the men's 10,000-m race at the 2000 Summer Olympic Games. After a race that lasted for more than 27 min, the differences between the gold and silver

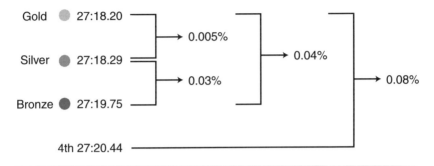

Figure I.1 The difference in performance (%) between medalists in the men's 10,000 m at the 2000 Sydney Olympics.

medalists and the silver and bronze medalists were an incredible 0.005% and 0.03%, respectively! These results help to illustrate why many elite endurance athletes use altitude as part of their overall training. Athletes believe that altitude training may provide them with that critical 0.5% edge in performance, which may be the difference between winning and not winning a medal.

1968 Summer Olympic Games—Mexico City

As already noted, most of the current interest in altitude training and its effect on human athletic performance originated with the 1968 Summer Olympic Games held in Mexico City at an elevation of 2,300 m (7,544 ft). Going into the 1968 Olympics, many coaches, athletes, and sport scientists predicted that the environmental conditions of Mexico City would favor the sprints and jumps in track and field but would have a negative effect on the aerobically based long distance events. Essentially, this prediction held true. One of the most memorable performances of the Mexico City Olympics was turned in by American athlete Bob Beamon, who jumped 8.90 m (29 ft 2.5 in.) in the long jump, breaking the world record by an incredible 56 cm (22.0 in.) (figure I.2). Years later, biomechanical engineers calculated that about 15 cm (5.9 in.) of the 56-cm (22.0 in.) margin by which Beamon broke the long jump world record was due solely to altitude and its favorable effect on approach speed and reduction of aerodynamic drag during the aerial phase of the jump. The combined effect of altitude plus a favorable wind (measured at 2.0 m · sec^{-1}) accounted for approximately 31 cm (12.2 in.) of the 56-cm (22.0 in.) margin (Ward-Smith 1986). Thus, Beamon's incredible leap at the Mexico City Olympics broke the world record by a sizable 25 cm (9.8 in.) independent of any "altitude effect." In addi-

Figure I.2 American Bob Beamon setting a world record in the long jump at the 1968 Mexico City Olympics.

tion to Beamon's world record in the long jump, world records were set in the men's 100 m, 200 m, and 400 m, as well as in the 4 × 100-m relay and 4 × 400-m relay. World records were also established in the 400-m intermediate hurdles and triple jump. Collectively, world records were broken in 8 of the 14 sprint, jump, and weight events; and Olympic records were either tied or broken in five of the six events in which world records were not established (table I.1). The percentage of improvement over previous world records ranged from 0.3% in the 400 m to 6.6% in the long jump.

In contrast, no world records were set in the middle or long distance running events except by Australia's Ralph Doubell, who tied the world record in the 800-m run (table I.2). Gold medalists in the 1,500 m, 3,000-m steeplechase, 5,000 m, 10,000 m, and marathon ran slower by 0.8%, 5.3%, 6.1%, 6.5%, and 8.4%, respectively, compared with the 1968 world records in those events. The negative effect of Mexico City's altitude on native lowlanders was evident in the performance of Ron Clarke of Australia, who was the world-record holder in the 5,000-m and 10,000-m runs at the time of the 1968 Mexico City Olympics (figure I.3). Clarke finished fifth in the 5,000-m race, running about 1 min slower than his

Table I.1 Results of Sprint, Jump, and Weight Events in Men's Track and Field at the 1968 Mexico City Olympics

Event	Gold	Silver	Bronze
100 m	Hines (USA) 9.95 WR (previous WR 9.9*)	Miller (Jam) 10.04	Greene (USA) 10.07
200 m	Smith (USA) 19.83 WR (previous WR 19.92)	Norman (Aus) 20.06	Carlos (USA) 20.10
400 m	Evans (USA) 43.86 WR (previous WR 44.0)	James (USA) 43.97	Freeman (USA) 44.41
4 x 100 m	USA 38.24 WR (previous WR 38.6)	Cuba 38.40	France 38.43
4 x 400 m	USA 2:56.16 WR (previous WR 2:59.6)	Kenya 2:59.64	Germany 3:00.54
110-m high hurdles	Davenport (USA) 13.33 = OR (WR 13.2)	Hall (USA) 13.42	Ottoz (Ita) 13.46
400-m IM hurdles	Hemery (GB) 48.12 WR (previous WR 48.8)	Hennige (Ger) 49.02	Sherwood (GB) 49.03
Long jump	Beamon (USA) 8.90/29-2 1/2 WR (previous WR 8.35/27-4 3/4)	Beer (GDR) 8.19/26-10 1/2	Boston (USA) 8.16/26-9 1/4
Triple jump	Saneyev (USSR) 17.39/57-3/4 WR (previous WR 17.03/55-10 1/2)	Prudencio (Bra) 17.27/56-8	Gentile (Ita) 17.22/56-6
High jump	Fosbury (USA) 2.24/7-4 1/4 OR (WR 2.28/7-5 3/4)	Caruthers (USA) 2.22/7-3 1/4	Gavrilov (USSR) 2.20/7-2 1/2
Pole vault	Seagren (USA) 5.40/17-8 1/2 OR (WR 5.41/17-9)	Schiprowski (Ger) 5.40/17-8 1/2 OR	Nordwig (GDR) 5.40/17-8 1/2 OR

Event	Gold	Silver	Bronze
Shot put	Matson (USA) 20.54/67-4 3/4 (WR 21.78/71-5 1/2	Woods (USA) 20.12/66-1/4	Gushchin (USSR) 20.09/65-11
Discus	Oerter (USA) 64.78/212-6 OR (WR 68.40/224-5)	Milde (GDR) 63.08/206-11	Danek (Cze) 62.92/206-5
Hammer	Zsivotsky (Hun) 73.36/240-8 OR (WR 73.76/242-0)	Klim (USSR) 73.28/240-5	Lovasz (Hun) 69.78/228-11

Results for the sprint events are given in seconds except for the 4 × 400 m relay, which is given in minutes and seconds. Results for the jump and weight events are given in meters first and feet and inches second.

*Hand-timed performance (considered inferior to the electronically timed performance of Hines).

IM = intermediate; WR = world record; OR = Olympic record.

world record. A few days later he finished sixth in the 10,000-m race, running more than 2 min slower than his world record.

Another interesting trend seen at the 1968 Summer Olympics was the fact that several of the medalists in the men's middle and long distance events came from altitude-based countries such as Kenya and Ethiopia. Indeed, the 1968 Mexico City Olympics marked the first of several Olympic Games dominated by Kenyan and Ethiopian distance runners. In Mexico City, Kenyan athletes won 39% of the medals (7 of 18), including three gold medals, in events ranging from 800 m to the marathon. Kenya's Kipchoge Keino (figure I.4) established himself as one of the great distance runners of his era by winning the 1,500-m run and establishing an Olympic record in the process. Keino also finished a very close second to Tunisia's Mohamed Gammoudi in the 5,000-m run.

The Mexico City Olympics made it obvious that in order to compete successfully in endurance events at altitude, it was advantageous either to be a native of an altitude-based country such as Kenya or Ethiopia or to have completed extensive altitude training beforehand. Logically, the next question that many athletes, coaches, and sport scientists began asking was "What effect does living and/or training at altitude have on *sea level* performance?" As of 1968 there had been only a few scientific studies of this question, and those investigations were not well controlled. Thus, most of the early research on the effect of altitude training on athletic performance came about as a result of all the interest generated by the 1968 Mexico City Olympics.

Table I.2 Results of Middle and Long Distance Events in Men's Track and Field at the 1968 Mexico City Olympics

Event	Gold	Silver	Bronze
800 m	Doubell (Aus) 1:44.3 = WR	Kiprugut (Ken) 1:44.5	Farrell (USA) 1:45.4
1,500 m	Keino (Ken) 3:34.9 OR (WR 3:33.1)	Ryun (USA) 3:37.8	Tummler (Ger) 3:39.0
3,000-m steeplechase	Biwott (Ken) 8:51.0 (WR 8:24.2)	Kogo (Ken) 8:51.6	Young (USA) 8:51.8
5,000 m	Gammoudi (Tun) 14:05.0 (WR 13:16.6)	Keino (Ken) 14:05.2	Temu (Ken) 14:06.4
10,000 m	Temu (Ken) 29:27.4 (WR 27:39.4)	Wolde (Eth) 29:28.0	Gammoudi (Tun) 29:34.2
Marathon	Wolde (Eth) 2:20:26 (WB 2:09:36)	Kimihara (Jap) 2:23:31	Ryan (NZ) 2:23:45

Results are given in minutes and seconds except for the marathon, which is given in hours, minutes, and seconds.

WR = world record; OR = Olympic record; WB = world best (WR not recognized in the marathon).

Post-Mexico City

Since the 1968 Summer Olympics, several scientific studies have been conducted for the purpose of examining the effect of altitude training on athletic performance. In addition to evaluating the influence of altitude training on sea level performance, several studies have attempted to answer other important questions, including the following:

- What is the optimal altitude at which to train?
- How long does an athlete need to train at altitude in order to gain beneficial physiological effects?
- How long does the "altitude effect" last after return to sea level?
- Do athletes from altitude-based countries have an inherent competitive advantage over athletes from lowland countries?

Figure I.3 Australia's Ron Clarke (102) and Kenya's Naftali Temu (575) in the 10,000-m run at the 1968 Mexico City Olympics.

Figure I.4 Kenya's Kipchoge "Kip" Keino (right), gold medalist in the 1,500-m run at the 1968 Mexico City Olympics. Temmate Ben Jipcho congratulates Keino.

Chapter 3, "Performance at Sea Level Following Altitude Training," reviews and summarizes the scientifically based "answers" to these questions. In addition, chapter 5 ("Altitude Training Programs of Successful Coaches and Athletes") addresses these questions by presenting the views of several coaches and athletes who have used altitude training in preparation for international competition.

As described earlier, the 1968 Mexico City Olympics was the first of several Olympic Games in which the middle and long distance events in men's track and field were dominated by altitude-based countries from East Africa, in particular Kenya and Ethiopia. From the 1968 Mexico City Olympics through the 2000 Sydney Olympics, Kenyan and Ethiopian male athletes won a relatively large proportion of medals in middle and long distance running events (800 m-marathon) compared with the rest of the world (figure I.5). This trend reached a peak at the 2000 Olympic Games, in which Kenyan and Ethiopian athletes won 61% of the medals awarded in the distance events, including five of six gold medals! Exceptions to this trend occurred in 1976 when the Montreal Olympic Games were boycotted by the African nations in protest against South Africa's apartheid policy, and in 1980 and 1984 when the nations of Eastern Africa were experiencing political and economic instability. Kenyan and Ethiopian distance runners have also been extremely successful in recent years in breaking world records in several long distance events. Ethiopia's Haile Gebrselassie and Kenya's Paul Tergat (figure I.6) have

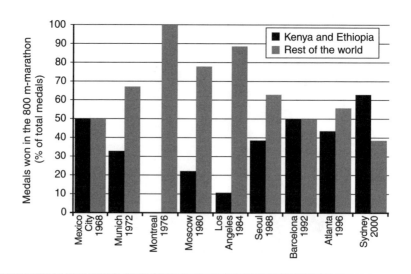

Figure I.5 Medals won by Kenyan and Ethiopian male athletes in distance running events (800 m-marathon) in the Olympic Games since 1968.

Figure I.6 Ethiopia's Haile Gebrselassie (left) and Kenya's Paul Tergat (right) compete in the 10,000-m run at the Sydney 2000 Olympics. The race was won by Gebrselassie in a photo finish.

each been consistently among the best in the world in the 10,000-m run. At present, Haile Gebrselassie holds the world record for the 5,000-m (12 min:39.36 sec) and 10,000-m runs (26:22.75). The success of the Kenyan and Ethiopian runners has helped promote a continuing interest in altitude training and its effect on athletic performance. Chapter 5 ("Altitude Training Programs of Successful Coaches and Athletes") provides more detail on the training methods of the Kenyan distance runners.

Present and Future

Several novel approaches and modalities are currently being used for altitude training. The "live high-train low" (LHTL) strategy of altitude training proposes that athletes can improve sea level endurance performance by living high (2,000-2,700 m/6,560-8,860 ft) while simultaneously training at low elevation (≤1,000 m/3,280 ft). It is believed that living at a relatively high altitude brings about increases in serum erythropoietin, red blood cell mass, and hemoglobin. Simultaneous training at low altitude allows athletes to work at intensities that are similar to those at sea level, thereby inducing beneficial peripheral and neuromuscular adaptations. In turn, the hematological and neuromuscular improvements that result from LHTL altitude training lead to the enhancement of sea level $\dot{V}O_2$max and endurance performance (Levine and Stray-Gundersen 1997).

A modification of LHTL altitude training is currently being used in Finland, a country that does not have highlands or mountain ranges. The Finns have developed the "nitrogen house," a normobaric hypoxic apartment that is used to simulate an altitude living environment. Another modification of the LHTL strategy involves the use of supplemental oxygen for the purpose of simulating sea level environmental conditions during high-intensity workouts conducted at altitude. Supplemental oxygen training is currently used effectively at the U.S. Olympic Training Center in Colorado Springs, Colorado, where U.S. national team athletes live at 1,860 m (6,100 ft) or higher, but can train at "sea level" with the aid of supplemental oxygen. Finally, "simulated altitude" devices such as the Colorado Altitude Room™ and the Hypoxico Altitude Tent™ are utilized by athletes to "sleep high" and train low. These novel training strategies and devices are described in detail in chapter 6, "Current Practices and Trends in Altitude Training."

References

Hopkins, W.G., J.A. Hawley, and L.M. Burke. 1999. Design and analysis of research on sport performance enhancement. *Medicine and Science in Sports and Exercise* 31: 472-485.

Levine, B.D., and J. Stray-Gundersen. 1997. "Living high-training low": effect of moderate-altitude acclimatization with low-altitude training on performance. *Journal of Applied Physiology* 83: 102-112.

Ward-Smith, A.J. 1986. Altitude and wind effects on long jump performance with particular reference to the world record established by Bob Beamon. *Journal of Sport Sciences* 4: 89-99.

Part I

The Science of Altitude Training

Chapter 1
POTENTIAL PHYSIOLOGICAL BENEFITS OF ALTITUDE TRAINING

Many athletes use altitude training for the purpose of enhancing performance. Most of these athletes participate in endurance-based sports such as distance running, triathlon, road cycling, and cross-country skiing. Each season they spend considerable time and resources training at altitude in order to improve their workout capabilities and competitive performance. Why do many athletes consider altitude training an important part of their overall training program? In considering this question, it is important to understand the "science" behind altitude training. Different aspects of human physiology are affected in different ways at altitude. In general, the various systems of the human body—pulmonary, cardiovascular, endocrine, skeletal muscle—respond and adjust in an effort to provide enough oxygen to survive in the hypoxic environment of high altitude. Some of these life-sustaining physiological responses may also enhance athletic performance, particularly in endurance sports.

Hematological

The ability to utilize oxygen for energy production is an important physiological factor for success in endurance sports. This ability is known more formally as an athlete's *oxygen consumption*. Oxygen consumption is expressed quantitatively as $\dot{V}O_2$, which stands for the volume of oxygen consumed per minute and utilized for aerobic energy production. The term $\dot{V}O_2$ can be expressed in absolute (liters of oxygen $\cdot$ min^{-1} [L $\cdot$ min^{-1}]) or relative units (milliliters of oxygen $\cdot$ kilogram of body weight^{-1} $\cdot$ min^{-1} [ml $\cdot$ kg^{-1} $\cdot$ min^{-1}]). During exhaustive maximal aerobic exercise, oxygen consumption is referred to as $\dot{V}O_2$max and can rise to levels of 65 to 75 ml $\cdot$ kg^{-1} $\cdot$ min^{-1} and 75 to 85 ml $\cdot$ kg^{-1} $\cdot$ min^{-1} in well-trained female and male aerobic athletes, respectively. By comparison, typical values for untrained females and males may range from 35 to 40 ml $\cdot$ kg^{-1} $\cdot$ min^{-1} and 45 to 50 ml $\cdot$ kg^{-1} $\cdot$ min^{-1}, respectively.

It is important to understand that an athlete's $\dot{V}O_2$ is determined by both central and peripheral physiological factors. Central factors influence the rate of *oxygen delivery* from the heart to the working muscles via the blood, whereas peripheral factors control the rate of *oxygen extraction* from the blood at the site of the working muscles. Quantitatively, this relationship is expressed in the Fick equation:

$$\dot{V}O_2 = O_2 \text{ delivery} \times O_2 \text{ extraction}$$
$$= \text{cardiac output} \times \text{arterial-venous } O_2 \text{ difference}$$

Cardiac output ($\dot{Q}$) is the volume of blood ejected by the left ventricle of the heart per minute. As shown in figure 1.1, cardiac output is a function of *heart rate (HR)* and *stroke volume (SV)*. Heart rate is simply the number of heartbeats per minute, whereas stroke volume is the volume of blood ejected by the left ventricle of the heart in a single heartbeat. By calculating the product of heart rate (beats · min^{-1}) and stroke volume (ml blood · heartbeat^{-1}), one can determine cardiac output (L blood · min^{-1}) and thus the volume of blood that is ejected by the heart, a percentage of which is delivered to the working muscles. The *arterial-venous O_2 difference* (a-$\bar{v}$ O_2 diff) is defined as the difference between the oxygen concentration of "oxygen-enriched" arterial blood circulating from the heart to the working muscles and the oxygen concentration of "oxygen-reduced" venous blood circulating from the working muscles back to the heart (figure 1.1). Thus, the a-$\bar{v}$ O_2 difference provides an indirect measure of the amount of oxygen extracted by the working

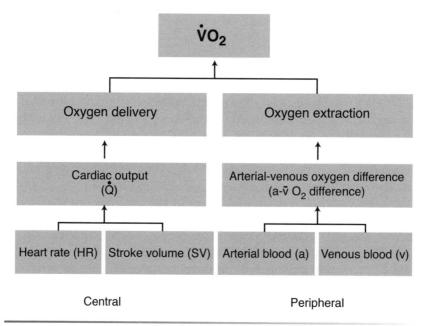

Figure 1.1 Central and peripheral factors affecting oxygen consumption ($\dot{V}O_2$).

muscles and utilized for aerobic energy production. According to the Fick equation, the product of cardiac output (L blood · min^{-1}) and a-$\bar{v}$ O$_2$ difference (ml O$_2$ · dL blood $^{-1}$) is equivalent to the absolute (L̇O$_2$ · min^{-1}) or relative V̇O$_2$ (ml O$_2$ · kg^{-1} · min^{-1}) and thus an individual's oxygen consumption.

As just described, cardiac output is the volume of blood ejected by the left ventricle of the heart per minute. A significant portion of the blood is made up of red blood cells (RBCs) or erythrocytes. The erythrocyte portion of the blood is referred to as the *hematocrit (Hct)*, which is expressed as a percentage of the number of erythrocytes relative to total blood volume. Hematocrits for healthy individuals residing at low elevation range from 35% to 45% for women and 40% to 50% for men. An individual's total blood volume contains trillions of erythrocytes. In turn, a single erythrocyte contains about 250 million molecules of *hemoglobin (Hb)*. Hemoglobin levels for healthy individuals residing at low elevation range from 12 to 16 grams of hemoglobin per deciliter of blood (g · dL^{-1}) for women, and 13 to 18 g · dL^{-1} for men. The primary physiological function of hemoglobin is to transport oxygen from the lungs to the body's organs and tissues.

As shown in figure 1.2, each molecule of hemoglobin consists of four identical subunits, which are classified as two alpha (α) and two beta (β)

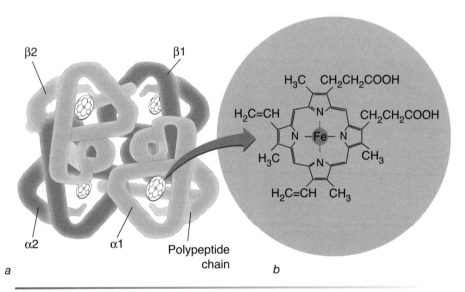

Figure 1.2 Structure of hemoglobin: *(a)* Hemoglobin; *(b)* iron-containing heme group.

Fig. 18.4, p. 582 from *Human Anatomy and Physiology*, 2nd ed. by Elaine N. Marieb. Copyright © 1992 by the Benjamin/Cummings Publishing Company, Inc. Reprinted by permission of Pearson Education, Inc.

subunits. Each of these subunits is composed of a helical polypeptide structure that supports a polyporphyrin ring. At the center of the polyporphyrin ring is an atom of iron (Fe) that serves as the exclusive site for oxygen binding and release. Thus, a single molecule of hemoglobin has the capacity of binding and transporting four molecules of oxygen, and a single RBC has the capacity to transport approximately 1 billion molecules of oxygen. Quantitatively, each gram of hemoglobin has the capacity to transport 1.34 ml of oxygen. An individual's *arterial oxygen content* (C_aO_2) can be estimated using the hemoglobin-oxygen constant (1.34 ml $O_2 \cdot$ g Hb^{-1}) and hemoglobin concentration (g Hb $\cdot$ dL^{-1}). Assuming an average hemoglobin concentration of 13 and 15 g $\cdot$ dL^{-1} for healthy women and men, respectively, C_aO_2 is equivalent to 17.7 and 20.4 ml $O_2 \cdot dL^{-1}$. Calculation of C_aO_2 also assumes normal arterial oxyhemoglobin saturation (97%-98%) and approximately 0.3 ml $O_2 \cdot dL^{-1}$ dissolved in the blood.

The process of oxygen transport is regulated by changes in the *partial pressure of oxygen* that take place from the moment we inhale oxygen through our nose and mouth until it reaches our body's tissues and organs. The *partial pressure of inspired oxygen* (P_IO_2) is determined by the ambient barometric pressure and O_2 concentration of the inspired air. For example, at sea level the concentration of O_2 is approximately 20.93% and the barometric pressure is 760 mm Hg. This results in a P_IO_2 of approximately 149 mm Hg determined as follows:

$$P_IO_2 = (760 \text{ mm Hg} - 47 \text{ mm Hg}) \times 0.2093 = 149 \text{ mm Hg}$$

where 760 mm Hg represents the barometric pressure at sea level, 47 mm Hg represents the pressure of water vapor in the lungs, and 0.2093 is the concentration of oxygen at sea level expressed as a *fraction of inspired oxygen* (F_IO_2).

The partial pressure of oxygen decreases as the inspired air moves from the nose and mouth to the lungs (figure 1.3). Thus, the *partial pressure of oxygen at the alveolar level of the lungs* (P_AO_2) is approximately 105 mm Hg. Blood entering the lungs via the pulmonary arteries contains erythrocytes whose oxygen concentration is relatively low. In other words, the *partial pressure of oxygen of arterial blood* (P_aO_2) entering the lungs is approximately 40 mm Hg. This pressure difference (105 vs. 40 mm Hg), or gradient, favors the diffusion of oxygen molecules from the alveoli of the lungs to the pulmonary blood where oxygen binds to available sites on hemoglobin molecules, a process that takes about 0.75 sec. As a result, "oxygen-enriched" blood exits the lungs with a P_aO_2 of 100 mm Hg and is transported via the pulmonary veins to the left ventricle of the heart, from which it is then circulated throughout the body. When oxygen-enriched arterial blood arrives at the capillary bed

Potential Physiological Benefits 7

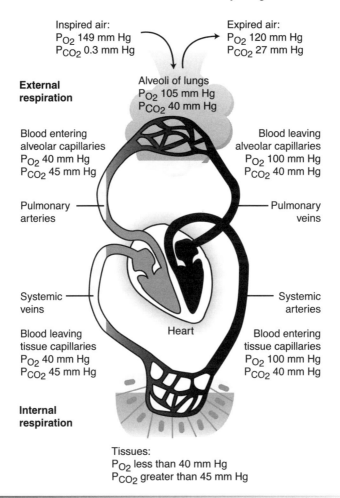

Figure 1.3 Partial pressure gradients promoting gas movements in the body. Gradients promoting oxygen and carbon dioxide exchanges across the respiratory membrane in the lungs (external respiration) are shown in the top part of the figure. Gradients promoting gas movements across systemic capillary membranes in the body tissues (internal respiration) are indicated in the bottom part.

Fig. 23.15, p. 748 from *Human Anatomy and Physiology*, 2nd ed. by Elaine N. Marieb. Copyright © 1992 by the Benjamin/Cummings Publishing Company, Inc. Reprinted by permission of Pearson Education, Inc.

of a skeletal muscle, the pressure gradient favors the release of oxygen from hemoglobin (P_aO_2 ~100 mm Hg) to the muscle (P_aO_2 ~30 mm Hg), where it will be utilized for aerobic energy production. The blood exits the capillary bed of the muscle in an "oxygen-reduced" state (P_vO_2 ~40 mm Hg) and returns to the right ventricle of the heart to repeat the process of oxygenation in the lungs.

The scientific rationale for using altitude training to enhance aerobic performance is based on the body's response to changes in P_IO_2 and P_aO_2. Figure 1.4 shows the changes that take place in the partial pressure of oxygen as one ascends from sea level to the top of Mount Everest (8,852 m/29,035 ft). P_IO_2 at sea level is equal to 149 mm Hg. P_IO_2 at Mexico City (2,300 m/7,544 ft) drops to approximately 123 mm Hg, whereas at the summit of Mount Everest P_IO_2 is approximately 50 mm Hg, only about 30% of sea level P_IO_2.

As a result of the altitude-induced decrease in P_IO_2, there is a simultaneous decrease in P_aO_2. This altitude-induced decrement in systemic P_aO_2 leads to a drop in kidney P_aO_2 and kidney oxygenation (Ou et al. 1998; Richalet et al. 1994). It is hypothesized that this reduction in kidney oxygenation stimulates the synthesis and release *of erythropoietin (EPO)* (Porter and Goldberg 1994; Richalet et al. 1994), the principal hormone that regulates mammalian erythrocyte and hemoglobin production. In turn, an increase in serum EPO concentration stimulates *erythropoiesis*, the production of new RBCs, in the red bone marrow (figure 1.5) by enhancing the mitotic frequency of erythroid precursors, specifically the colony forming unit-erythroid (CFU-E). Erythropoietin receptors are present on the surface of CFU-E. Binding of EPO to CFU-E receptors initiates the production of cellular transcription factors, synthesis of membrane and cytoskeletal proteins, synthesis of heme and hemo-

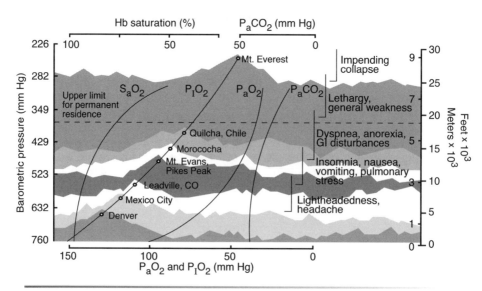

Figure 1.4 Changes in barometric pressure and the partial pressure of oxygen upon ascent from sea level to the summit of Mount Everest (8,852 m/29,035 ft).

Reprinted, by permission, from W.D. McArdle, F.I. Katch, and V.L. Katch, 1991, *Exercise physiology*, 3rd ed. (Philadelphia: Lea and Febiger), 530.

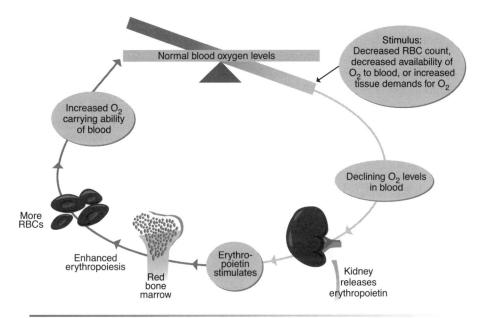

Figure 1.5 The erythropoietin (EPO) mechanism.

Fig. 18.6, p. 584 from *Human Anatomy and Physiology*, 2nd ed. by Elaine N. Marieb. Copyright © 1992 by the Benjamin/Cummings Publishing Company, Inc. Reprinted by permission of Pearson Education, Inc.

globin, and the terminal differentiation of cells (Bell 1996). The erythrocyte maturation process takes approximately five to seven days from the initial altitude-induced increase in serum EPO (Bell 1996; Flaharty et al. 1990). A physiological threshold for increasing RBC mass and hemoglobin has been described to be at a P_aO_2 of approximately 70 mm Hg, which is equivalent to approximately 1,600 m (5,250 ft) (Weil et al. 1968). However, more current research has shown that there is great individual variability in the altitude-induced erythropoietic response (Chapman et al. 1998).

On the basis of the previously described scientific rationale, the potential physiological benefits of altitude training can be summarized as follows. Exposure to altitude results in decrements in P_IO_2 and P_aO_2. Reductions in kidney P_aO_2 and kidney tissue oxygenation stimulate the synthesis and release of EPO, which subsequently leads to an increase in RBC mass and hemoglobin concentration. These hematological changes may significantly improve an athlete's $\dot{V}O_2$ by enhancing the blood's ability to deliver oxygen to the exercising muscles. It has been shown that these improvements in RBC mass, hemoglobin concentration, and $\dot{V}O_2$ enhance aerobic performance. Essentially, altitude training is viewed by many athletes and coaches as a "natural" or "legal" method of blood doping.

The scientific rationale for altitude training is supported by data from laboratory studies that have used artificial methods to increase EPO, RBC mass, and hemoglobin concentration and subsequently produced significant improvements in $\dot{V}O_2$max and endurance performance. Ekblom and Berglund (1991) studied the influence of DNA-recombinant human erythropoietin (rhEPO) on hematological indexes and aerobic performance in healthy male physical education students ($\dot{V}O_2$max = 4.52 L · min^{-1}). The subjects received an rhEPO dosage of 20 international units (U) per kilogram of body weight (BW) administered subcutaneously three times per week for six weeks (calculated daily average was equivalent to 8.6 U · kg^{-1} BW per day). Following treatment, average hematocrit (pre = 44.5%, post = 49.7%) and hemoglobin concentration (pre = 15.2, post = 16.9 g · dL^{-1}) increased significantly by 12% and 11%, respectively, compared with preexperimental values (figure 1.6).

In terms of aerobic performance, treadmill $\dot{V}O_2$max increased 8% ($p < 0.05$) from 4.52 to 4.88 L · min^{-1} following rhEPO treatment (figure 1.6). Furthermore, this increase in maximal oxygen consumption was observed in all 15 subjects. As part of an effort to study the longitudinal effect of rhEPO supplementation on $\dot{V}O_2$max, the subjects reported back to the laboratory after one and three weeks following the rhEPO treatment. Results of the follow-up tests indicated that the marked increase in $\dot{V}O_2$max observed subsequent to six weeks of rhEPO administration

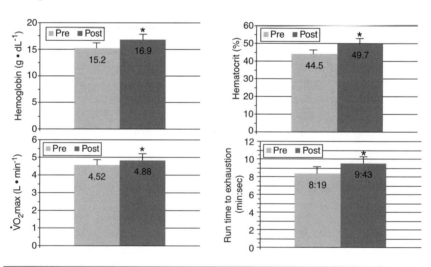

Figure 1.6 The effect of recombinant human erythropoietin (rhEPO) administration on hemoglobin, hematocrit, maximal oxygen consumption ($\dot{V}O_2$max), and endurance performance. * Significantly different vs. pre ($p < 0.05$).

Adapted, by permission, from B. Ekblom and B. Berglund, 1991, "Effects of erythropoietin administration on maximal aerobic power," *Scandinavian Journal of Medicine and Science in Sports* 1: 90.

had diminished in most subjects within 14 days of cessation of rhEPO. In addition to $\dot{V}O_2$max, significant improvements were observed in running performance following six weeks of rhEPO treatment. Running performance (figure 1.6), quantified as total run time for an incremental treadmill test (13-16 km · hr^{-1} with a 1° increment every 2 min), increased 17% (p < 0.05) from 8 min:19 sec to 9 min:43 sec. A significant increase in systolic blood pressure and a significant decrease in heart rate were observed during submaximal exercise (6 min of cycling at 200 W) subsequent to the six weeks of rhEPO treatment (Berglund and Ekblom 1991). Collectively, the results of Ekblom and Berglund (1991) suggested that moderate rhEPO administration results in a marked increase in RBC mass and hemoglobin concentration, as well as significant improvements in aerobic performance.

More recently, Birkeland et al. (2000) evaluated the effect of rhEPO administration on $\dot{V}O_2$max in 10 trained male endurance athletes. Subjects received a total dosage of 5,000 U rhEPO, administered subcutaneously in three separate injections per week for four weeks (calculated daily average equivalent to 25.9-33.1 U · kg^{-1} BW per day), which was a higher dosage compared with that used in previous studies. Hematocrit increased 19% (p < 0.05) from a baseline value of 42.7% to a maximum value of 50.8% when measured on the day after the end of the rhEPO treatment period. Maximal oxygen uptake was measured on a cycle ergometer prior to, as well as during, a four-week period following the cessation of rhEPO administration. Pre-rhEPO $\dot{V}O_2$max was equivalent to 63.6 ml · kg^{-1} · min^{-1} and increased by 7% (p < 0.05), 6% (p < 0.05), 6% (no statistically significant difference; NSD), 5% (p < 0.05), and 3% (NSD) when measured 2, 7, 14, 21, and 28 days, respectively, following the end of the 30-day rhEPO administration period. There were no significant changes in hematocrit or $\dot{V}O_2$max in a fitness-matched control group over the same period of time. Additional studies on the performance effects of rhEPO administration are summarized in table 1.1. Collectively, the data from these studies suggest that moderate rhEPO supplementation results in significant improvements in hemoglobin (9-14%), hematocrit (11-19% relative increase), $\dot{V}O_2$max (7-9%), and endurance performance (17%).

Research by Chapman et al. (1998) suggests that some athletes experience a better hematological response at altitude than others. Female and male collegiate runners who completed either "live high-train low" (LHTL) or traditional live high-train high altitude training were classified as "responders" or "non-responders" based on their performance in a postaltitude 5-km run. On average, responders demonstrated a significant 4% improvement (37 sec) in the postaltitude 5-km run versus their prealtitude performance, whereas the nonresponders were approximately 1% slower (14 sec). Hematological data showed that

Table 1.1 Effects of Recombinant Human Erythropoietin Administration

Author	Hct (% increase)	Hb (% increase)	$\dot{V}O_2$max (% increase)	Endurance performance (% improvement)
Audran et al. 1999	11*	9*	9*	NR
Balsom et al. 1994	NR	11*	8*	NR
Birkeland et al. 2000	19*	14*	7*	NR
Ekblom and Berglund 1991	12*	11*	8*	17*

Changes in hematocrit are reported as relative percent increase.
*Significantly greater ($p < 0.05$) versus pretreatment values; Hct = hematocrit (%); Hb = hemoglobin (g · dl^{-1}); NR = not reported or measured; $\dot{V}O_2$max = maximal oxygen consumption.

the responders had a significantly larger increase in serum EPO (52%) compared with the nonresponders, who demonstrated a 34% increase in serum EPO. Similarly, postaltitude RBC mass for the responders was 8% higher ($p < 0.05$), but the nonresponders' RBC mass was only 1% higher (not statistically significant) compared with prealtitude values. A breakdown of responders indicated that 82% came from the LHTL group, whereas 18% came from the live high-train high group. The authors concluded that each athlete may need to follow an altitude training program that places him or her at an individualized, optimal altitude for living and another altitude for training, thereby producing the best possible hematological response.

We have now seen that several laboratory-based studies have provided data in support of the scientific rationale for altitude training as a method for improving maximal oxygen consumption and endurance performance. This rationale is based on the fact that exposure to altitude produces an increase in serum EPO that subsequently leads to increments in erythrocyte and hemoglobin concentration. Many coaches and athletes believe that these hematological changes significantly improve an athlete's $\dot{V}O_2$ by enhancing the blood's capacity to deliver oxygen to the exercising muscles. In support of this, the studies of Ekblom and Berglund (1991), Birkeland et al. (2000), and others demonstrated that enhanced serum EPO concentration via subcutaneous injection of rhEPO led to increases in hematocrit and hemoglobin, which subsequently resulted in improvements in $\dot{V}O_2$max and running performance. Given that the International Olympic Committee banned the use of all forms of blood doping in 1990 (Mottram 1999), altitude training is used by many athletes as a "legal" method of inducing hematological changes that appear to have a significant ergogenic effect on aerobic performance.

Skeletal Muscle

As described in the previous section, the primary reason endurance athletes train at altitude is to increase their RBC mass and hemoglobin concentration. In addition, there may be secondary physiological benefits to be gained as a result of altitude exposure (Mathieu-Costello 2001). For example, altitude training has been shown to increase skeletal muscle capillarity (Desplanches et al. 1993; Mizuno et al. 1990). In theory, this physiological adaptation enhances the exercising muscles' ability to extract oxygen from the blood.

A number of favorable physiological changes have also been found to occur within skeletal muscle microstructure as a result of altitude training. These changes include increased concentrations of myoglobin (Terrados et al. 1990), increased mitochondrial oxidative enzyme activity (Terrados et al. 1990), and a greater number of mitochondria (Desplanches et al. 1993), all of which serve to enhance the rate of oxygen utilization and aerobic energy production.

However, it should be noted that scientific data in support of altitude-induced skeletal muscle adaptations, particularly among well-trained athletes, are minimal. Only one of the previously cited investigations was conducted on elite athletes (Mizuno et al. 1990), whereas the other studies examined the effect of altitude training on the skeletal muscle characteristics of untrained individuals (Desplanches et al. 1993; Terrados et al. 1990). Additional studies with elite athletes have failed to demonstrate significant changes in skeletal muscle microstructure as a result of altitude training (Saltin et al. 1995; Terrados et al. 1988). Furthermore, one of the aforementioned studies (Desplanches et al. 1993) was conducted at simulated elevations (4,100-5,700 m/13,450-18,695 ft) that are impractical, that is, too high for athletes to train at. Thus, on the basis of the current scientific literature it is unclear whether altitude training, as practiced by most elite athletes at moderate elevations of 1,830 m to 3,050 m (6,000-10,000 ft), improves oxygen extraction and utilization via favorable changes in skeletal muscle capillarity, myoglobin, mitochondrial oxidative enzyme activity, and mitochondrial density. Additional research is warranted in order to answer these questions.

Another important physiological adaptation that may occur as a result of exposure to moderate altitude is an improvement in the capacity of the skeletal muscle and blood to buffer the concentration of hydrogen ions (H^+). During exercise, high concentrations of H^+ contribute to skeletal muscle fatigue by impairing actin-myosin cross-bridge cycling, reducing the sensitivity of troponin for calcium (Ca^{2+}), and inhibiting the enzyme phosphofructokinase (PFK), thereby reducing anaerobic energy production via glycolysis (McComas 1996). Thus, an enhanced H^+-buffering capacity may have a beneficial effect on athletic performance.

In support of this, Mizuno et al. (1990) reported a significant 6% increase in the buffering capacity of the gastrocnemius muscle of elite male cross-country skiers who lived at 2,100 m (6,890 ft) and trained at 2,700 m (8,860 ft) for 14 days. Significant improvements in maximal O_2 deficit (29%) and treadmill run time to exhaustion (17%) were observed after the athletes returned to sea level. In addition, a positive correlation ($r = 0.91$, $p < 0.05$) was demonstrated between the relative increase in buffering capacity of the gastrocnemius muscle and treadmill run time to exhaustion.

Gore et al. (2001) recently reported that skeletal muscle buffer capacity increased 18% ($p < 0.05$) in male triathletes, cyclists, and cross-country skiers following 23 days of living at 3,000 m (9,840 ft) and training at 600 m (1,970 ft). Furthermore, the researchers found that mechanical efficiency was significantly improved during a 4 × 4-min submaximal cycling test following the 23-day LHTL period.

The precise mechanisms responsible for an enhanced muscle buffering capacity following altitude training are unclear but may be related to changes in creatine phosphate (CP), muscle protein concentrations, or both (Mizuno et al. 1990). Improvements in blood buffering capacity may be due to increases in bicarbonate (Nummela and Rusko 2000), hemoglobin concentration, or both.

Genetics and Altitude Training

Recent advances in biological and molecular imaging technology have allowed scientists to identify the entire human genetic code. Known as the Human Genome Project, this multibillion-dollar effort involved thousands of international scientists who mapped approximately 100,000 genes from the 23 pairs of human chromosomes and identified their entire sequence. This incredible scientific accomplishment has provided us with vital information, particularly in the field of preventative medicine. The importance of genetic data is also evident in nonmedical areas as evidenced by a recent special report, published by the American College of Sports Medicine (ACSM) (Rankinen et al. 2001), that describes the human gene map for performance and health-related fitness phenotypes. Performance phenotypes that the ACSM paper addresses include cardiorespiratory endurance and muscular strength; health-related phenotypes include hemodynamic traits, anthropometry and body composition, insulin and glucose metabolism, and lipid profile.

In terms of altitude, observed differences in hematological, muscular, pulmonary, cardiovascular, and hormonal responses between high-altitude natives (Tibetans and Nepalese of the Himalaya Mountains; Peruvians of the Andes Mountains) and lowlanders suggest that genetics play an important role (Moore et al. 2002; Ramirez et al. 1999). More relevant to the focus of this book is the role of genetics as it relates to altitude training by athletes. An earlier section of this chapter described the physiological mechanism by which the hypoxic environment of high altitude stimulates the release of EPO from the kidneys, which subsequently increases the production of RBCs. It is now well accepted that the *hypoxia-inducible factor 1α* (HIF-1α) complex serves as the genetic regulator for the production and release of EPO from the kidneys (Caro 2001; Prabhakar 2001; Samaja 2001). The HIF-1α complex is located on human chromosome 14 and belongs to a class of genetic factors that regulate deoxyribonucleic acid (DNA) transcription into messenger ribonucleic acid (mRNA) (Samaja 2001). At altitude, the level of HIF-1α increases, and this in turn enhances the transcription of DNA to mRNA on the EPO gene (Vogt et al. 2001). The result is increased EPO production and release by the kidneys. In addition to regulating EPO, the HIF-1α complex modulates other physiological responses at altitude including glucose transport, glycolytic enzyme activity, inflammatory responses, and bone metabolism (Clerici and Matthay 2000; Gross et al. 2001; Samaja 2001).

Preliminary data suggest that a specific section of the EPO gene may discriminate between moderate versus significant increases in EPO among athletes at altitude. Athletes identified as having this specific section of the EPO gene, called an allele, had a 135% increase in serum EPO after 24 hr of simulated altitude at 2,800 m (9,186 ft). By comparison, athletes who did not have the specific allele of the EPO gene had a significantly lower 78% increase in serum EPO following 24 hr of simulated altitude (Witkowski et al. 2002). Genetic research as it relates to altitude training is a rapidly expanding area of study that will continue to provide us with intriguing but potentially controversial findings.

Summary

The scientific rationale for using altitude training as a method of improving aerobic performance is based on the fact that altitude-induced decrements in P_IO_2 and P_aO_2 stimulate the release of EPO, which subsequently leads to increments in erythrocyte and hemoglobin concentration. Many coaches and athletes believe that these hematological changes significantly improve an athlete's $\dot{V}O_2$ by enhancing the blood's capacity to deliver oxygen to the exercising muscles. In support of this, several scientific studies demonstrated that enhanced serum EPO concentration via subcutaneous injection of rhEPO led to increases in hematocrit and hemoglobin, which subsequently resulted in improvements in $\dot{V}O_2$max and running performance. Given that in 1990 the International Olympic Committee banned the use of all forms of blood doping, altitude training is used by many athletes as a "legal" method of inducing hematological changes that appear to have a significant ergogenic effect on aerobic performance.

In addition to positive hematological adaptations, a number of favorable physiological changes may also occur within skeletal muscle micro-

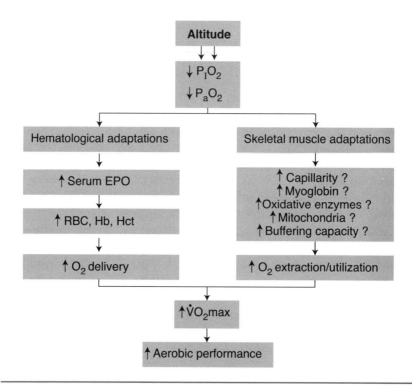

Figure 1.7 Summary of the purported physiological benefits of using altitude training for enhancement of aerobic performance. ↑ = increased or enhanced; ↓ = decreased; ? = potential adaptation (research has not shown conclusively that this physiological adaptation takes place as a result of altitude training); EPO = erythropoietin; Hb = hemoglobin; Hct = hematocrit; P_aO_2 = partial pressure of oxygen in arterial blood; P_IO_2 = partial pressure of inspired oxygen; RBC = red blood cells; $\dot{V}O_2$max = maximal oxygen consumption.

structure as a result of altitude training. These changes include enhanced capillarity as well as increased concentrations of myoglobin, increased mitochondrial oxidative enzyme activity, and a greater number of mitochondria, all of which serve to enhance the rate of aerobic energy production. However, it should be noted that scientific data in support of these skeletal muscle adaptations are minimal, particularly among well-trained athletes. Thus, it is difficult to say whether altitude training leads to improvements in oxygen extraction and utilization via favorable changes in skeletal muscle capillarity, myoglobin, mitochondrial oxidative enzyme activity, and mitochondrial density. One important skeletal muscle adaptation that may occur as a result of exposure to moderate altitude is an improvement in the capacity of the skeletal muscle to buffer the concentration of H^+. An enhanced H^+-buffering capacity may have a beneficial effect on aerobic and anaerobic performance by delaying skeletal muscle fatigue. The purported physiological benefits of using altitude training for the enhancement of aerobic performance are summarized in figure 1.7.

Identification of the human genetic code has provided fascinating new information regarding the role of genetics in altitude training. The role of the HIF-1α complex in the regulation of the EPO gene has been clearly demonstrated. Preliminary data suggest that some individuals may be genetically predisposed to responding better at altitude than others. This is a fertile area of research that will undoubtedly provide us with exciting but potentially controversial information in the near future.

References

Audran, M., R. Gareau, S. Matecki, F. Durand, C. Chenard, M-T. Sicart, B. Marion, and F. Bressolle. 1999. Effects of erythropoietin administration in training athletes and possible indirect detection in doping control. *Medicine and Science in Sports and Exercise* 31: 639-645.

Balsom, P.D., B. Ekblom, and B. Sjodin. 1994. Enhanced oxygen availability during high intensity intermittent exercise decreases anaerobic metabolite concentrations in blood. *Acta Physiologica Scandinavica* 150: 455-456.

Bell, C. 1996. Morphology of the erythron. In *Hematology: clinical and laboratory practice I*, edited by R.L. Bick, pp. 163-183. St. Louis: Mosby.

Berglund, B., and B. Ekblom. 1991. Effect of recombinant human erythropoietin treatment on blood pressure and some haematological parameters in healthy men. *Journal of Internal Medicine* 229: 125-130.

Birkeland, K.I., J. Stray-Gundersen, P. Hemmersbach, J. Hallen, E. Haug, and R. Bahr. 2000. Effect of rhEPO administration on serum levels of sTfR and cycling performance. *Medicine and Science in Sports and Exercise* 32: 1238-1243.

Caro, J. 2001. Hypoxia regulation of gene transcription. *High Altitude Medicine and Biology* 2: 145-154.

Chapman, R.F., J. Stray-Gundersen, and B.D. Levine. 1998. Individual variation in response to altitude training. *Journal of Applied Physiology* 85: 1448-1456.

Clerici, C., and M.A. Matthay. 2000. Hypoxia regulates gene expression of alveolar epithelial transport proteins. *Journal of Applied Physiology* 88: 1890-1896.

Desplanches, D., H. Hoppeler, M.T. Linoissier, C. Denis, H. Claasen, D. Dormois, J.R. Lacour, and A. Geyssant. 1993. Effects of training in normoxia and normobaric hypoxia on human muscle ultrastructure. *Pflugers Archiv* 425: 263-267.

Ekblom, B., and B. Berglund. 1991. Effects of erythropoietin administration on maximal aerobic power. *Scandinavian Journal of Medicine and Science in Sports* 1: 88-93.

Flaharty, K.K., J. Caro, A. Erslev, J.J. Whalen, E.M. Morris, T.D. Bjornsson, and P.H. Vlasses. 1990. Pharmacokinetics and erythropoietic response to human recombinant erythropoietin in healthy men. *Clinical Pharmacology and Therapeutics* 47: 557-564.

Gore, C.J., A.G. Hahn, R.J. Aughey, D.T. Martin, M.J. Ashenden, S.A. Clark, A.P. Garnham, A.D. Roberts, G.J. Slater, and M.J. McKenna. 2001. Live high:train low increases muscle buffer capacity and submaximal cycling efficiency. *Acta Physiologica Scandinavica* 173: 275-286.

Gross, T.S., N. Akeno, T.L. Clemens, S. Komarova, S. Srinivasan, D.A. Weimer, and S. Mayorov. 2001. Osteocytes upregulate HIF-1α in response to acute disuse and oxygen deprivation. *Journal of Applied Physiology* 90: 2514-2519.

Marieb, E.N. 1992. *Human anatomy and physiology* (2nd ed.). Redwood City, CA: Benjamin Cummings.

Mathieu-Costello, O. 2001. Muscle adaptation to altitude: Tissue capillarity and capacity for aerobic metabolism. *High Altitude Medicine and Biology* 2: 413-425.

McArdle, W.D., F.I. Katch, and V.L. Katch. 1991. *Exercise physiology*. Philadelphia: Lea and Febiger.

McComas, A.J. 1996. *Skeletal muscle form and function*. Champaign, IL: Human Kinetics.

Mizuno, M., C. Juel, T. Bro-Rasmussen, E. Mygind, B. Schibye, B. Rasmussin, and B. Saltin. 1990. Limb skeletal muscle adaptations in athletes after training at altitude. *Journal of Applied Physiology* 68: 496-502.

Moore, L.G., S. Zamudio, J. Zhuang, T. Droma, and R.V. Shohet. 2002. Analysis of the myoglobin gene in Tibetans living at high altitude. *High Altitude Medicine and Biology* 3: 39-47.

Mottram, D.R. 1999. Banned drugs in sport. *Sports Medicine* 27: 1-10.

Nummela, A., and H. Rusko. 2000. Acclimatization to altitude and normoxic training improve 400-m running performance at sea level. *Journal of Sports Science* 18: 411-419.

Ou, L.C., S. Salceda, S.J. Schuster, L.M. Dunnack, T. Brink-Johnsen, J. Chen, and J.C. Leiter. 1998. Polycythemic responses to hypoxia: Molecular and genetic mechanisms of chronic mountain sickness. *Journal of Applied Physiology* 84: 1242-1251.

Porter, D.L., and M.A. Goldberg. 1994. Physiology of erythropoietin production. *Seminars in Hematology* 31: 112-121.

Prabhakar, N.R. 2001. Oxygen sensing during intermittent hypoxia: Cellular and molecular mechanisms. *Journal of Applied Physiology* 90: 1986-1994.

Ramirez, G., P.A. Bittle, R. Rosen, H. Rabb, and D. Pineda. 1999. High altitude living: Genetic and environmental adaptation. *Aviation Space and Environmental Medicine* 70: 73-81.

Rankinen, T., L. Perusse, R. Rauramaa, M.A. Rivera, B. Wolfarth, and C. Bouchard. 2001. The human gene map for performance and health-related fitness phenotypes. *Medicine and Science in Sports and Exercise* 33: 855-867.

Richalet, J.P., J.C. Souberbielle, A.M. Antezana, M. Dechaux, J.L. Le Trong, A. Bienvenu, F. Daniel, C. Blanchot, and J. Zittoun. 1994. Control of erythropoiesis in humans during prolonged exposure to the altitude of 6,542 m. *American Journal of Physiology* 266: R756-R764.

Saltin, B., C.K. Kim, N. Terrados, H. Larsen, J. Svedenhag, and C.J. Rolf. 1995. Morphology, enzyme activities and buffer capacity in leg muscles of Kenyan and Scandinavian runners. *Scandinavian Journal of Medicine and Science in Sports* 5: 222-230.

Samaja, M. 2001. Hypoxia-dependent protein expression: Erythropoietin. *High Altitude Medicine and Biology* 2: 155-163.

Terrados, N., E. Jansson, C. Sylven, and L. Kaijser. 1990. Is hypoxia a stimulus for synthesis of oxidative enzymes and myoglobin? *Journal of Applied Physiology* 68: 2369-2372.

Terrados, N., J. Melichna, C. Sylven, E. Jansson, and L. Kaijser. 1988. Effects of training at simulated altitude on performance and muscle metabolic capacity in competitive road cyclists. *European Journal of Applied Physiology* 57: 203-209.

Vogt, M., A. Puntschart, J. Geiser, C. Zuleger, R. Billeter, and H. Hoppeler. 2001. Molecular adaptations in human skeletal muscle to endurance training under simulated hypoxic conditions. *Journal of Applied Physiology* 91: 173-182.

Weil, J.V., G. Jamieson, D.W. Brown, and R.F. Grover. 1968. The red cell mass-arterial oxygen relationship in normal man. *Journal of Clinical Investigation* 47: 1627-1639.

Witkowski, S., H. Chen, J. Stray-Gundersen, R.L. Ge, C. Alfrey, J.T. Prchal, and B.D. Levine. 2002. Genetic marker for the erythropoietic response to altitude. *Medicine and Science in Sports and Exercise* 34 (Suppl. 5): S246.

Chapter 2
PHYSIOLOGICAL RESPONSES AND LIMITATIONS AT ALTITUDE

In general, athletes experience greater physiological stress during competition and training at altitude compared with similar training at sea level. For example, a 10-km road race or an interval workout typically seems "a lot harder" for many athletes at altitude. The reason is that athletes experience a number of physiological changes upon exposure to altitude that may limit their ability to compete or train. Two of the most important physiological changes that occur are altitude-induced decrements in *arterial oxyhemoglobin saturation* and *maximal oxygen consumption*. These physiological limitations may force athletes to reduce their daily training volume, training intensity, or both and to modify their competition strategy from what they would normally do at sea level. In addition, exposure to altitude may bring about changes in *heart rate, hydration status, acid-base balance, carbohydrate utilization, iron metabolism, and immune function*. These physiological responses may force athletes to modify their normal sea level training regimen while at altitude in order to achieve an optimal training effect and avoid overreaching or overtraining. This chapter addresses each of these altitude-induced physiological changes in the context of athletic performance.

Arterial Oxyhemoglobin Saturation

As described in chapter 1, there is a decrease in the partial pressure of inspired oxygen (P_IO_2) upon exposure to altitude. In turn, this decrease in P_IO_2 leads to a reduction in the partial pressure of oxygen at the alveolar level of the lungs (P_AO_2), the site where oxygen diffuses through the pulmonary capillaries to the blood. A decrement in P_AO_2 ultimately leads to a reduction in the partial pressure of oxygen in arterial blood (P_aO_2), which results in fewer oxygen molecules binding to hemoglobin, that is, a decrease in *arterial oxyhemoglobin saturation (S_aO_2)*. Figure 2.1 shows the relationship between altitude and barometric pressure (P_B), P_IO_2, P_AO_2, P_aO_2, and S_aO_2 at several elevations: sea level; Leadville, Colorado (3,100 m/10,170 ft); Pikes Peak, Colorado (4,300 m/14,110 ft); and Mount Everest (8,852 m/29,035 ft).

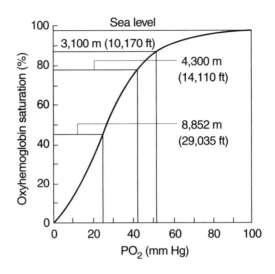

Oxygen Partial Pressures and Oxyhemoglobin Saturation at Various Altitudes

Pressures	Sea level	Leadville	Pikes Peak	Mt. Everest
Altitude (m/ft)	0	3,100/10,170	4,300/14,110	8,852/29,035
P_B (mm Hg)	760	530	440	250
P_IO_2 (mm Hg)	150	101	82	42
P_AO_2 (mm Hg)	105	71	47	30
P_aO_2 (mm Hg)	100	52	44	25
S_aO_2 (%)	96	90	82	48

Figure 2.1 Standard oxygen dissociation curve for blood pH = 7.4 and temperature = 37° C. Vertical lines indicate the partial pressure of oxygen (PO_2) and horizontal lines indicate percent hemoglobin saturation of arterial blood at sea level, 3,100 m, 4,300 m, and 8,852 m. P_B = barometric pressure; P_IO_2 = partial pressure of inspired oxygen; P_AO_2 = partial pressure of oxygen at the alveolar level of the lungs; P_aO_2 = partial pressure of oxygen in arterial blood; S_aO_2 = arterial oxyhemoglobin saturation.

Graph reprinted, by permission, from R.F. Grover, 1979, Physiological adaptations to high altitude. In *Sports medicine and physiology,* edited by R.H. Strauss (Philadelphia: W.B. Saunders), 333. Table reprinted, by permission, from E.M. Haymes and C.L. Wells, 1986, *Environment and human performance* (Champaign, IL: Human Kinetics), 73. Data from Grover 1978, Hannon 1978, Vogel and Hansen 1967, and West 1982.

At rest, P_aO_2 and S_aO_2 are slightly reduced at altitude (Banchero et al. 1966; Huang et al. 1984) despite a reflexive increase in pulmonary ventilation. During submaximal and maximal exercise at altitude, P_aO_2 and S_aO_2 are markedly lower compared with values during similar exercise at sea level (Gale et al. 1985; Hartley et al. 1973; Sylvester et al. 1981; Wagner et al. 1986). The reason for this effect may be that the altitude-induced reduction in P_AO_2 results in a decrease in pulmonary capillary diffusion time (Dempsey et al. 1984; Torre-Bueno et al. 1985; Wagner et al. 1986). In other words, because P_AO_2 is reduced at altitude, pulmonary capillary diffusion time is not long enough to allow for optimal oxygenation of the pulmonary arterial blood, thereby resulting in a lower S_aO_2.

Interestingly, this exercise-induced decrement in S_aO_2 appears to be more pronounced in well-trained athletes versus untrained individuals at both sea level and altitude. Lawler et al. (1988) initially reported this finding in a study that evaluated untrained and trained males who performed incremental cycle ergometer exercise at sea level and at a simulated altitude of 3,000 m (9,840 ft). Arterial oxyhemoglobin saturation was measured indirectly via pulse oximetry (S_pO_2), as opposed to direct measurement from arterial blood (S_aO_2). The trained group experienced a significantly greater decrement in S_pO_2 compared with the untrained group at both sea level (trained = 90.1%, untrained = 95.5%) and simulated altitude (trained = 77.3%, untrained = 86.3%), indicating that trained aerobic athletes experience greater impairment of arterial oxyhemoglobin saturation upon exposure to altitude than untrained individuals. It has been suggested that the reduced pulmonary capillary transit time experienced at altitude, in combination with the relatively high cardiac output, pulmonary blood flow, and hemoglobin content of endurance athletes, serves to widen the P_AO_2-P_aO_2 diffusion gradient, which results in a greater reduction in S_aO_2 in trained versus untrained individuals (Dempsey 1986; Gore et al. 1996; Torre-Bueno et al. 1985; Wagner et al. 1986).

Additional studies have provided further evidence that arterial oxyhemoglobin saturation is significantly reduced in endurance athletes during high-intensity exercise at altitude. Squires and Buskirk (1982) evaluated the effects of acute exposure to moderate hypoxia on S_pO_2 in male recreational runners. Each runner performed a maximal treadmill test in a hypobaric chamber at simulated altitudes of 362 m (1,190 ft), 914 m (3,000 ft), 1,219 m (4,000 ft), 1,524 m (5,000 ft), and 2,286 m (7,500 ft). Arterial oxyhemoglobin saturation during maximal treadmill exercise at 362 m averaged 91% and was significantly reduced to 87%, 87%, 84%, and 79% at 914 m, 1,219 m, 1,524 m, and 2,286 m, respectively (Squires and Buskirk 1982). Brosnan et al. (2000) reported that S_pO_2 was significantly lower at a simulated altitude of 2,100 m (6,890 ft) versus normoxia in Australian elite female road cyclists after each of three 10-min endurance cycling bouts ([1] 93% vs. 95%; [2] 93% vs. 96%; [3] 94% vs. 96%). Roberts et al. (1998) found that among male Australian national team cross-country skiers, S_aO_2 measured during maximal treadmill cross-country ski exercise was significantly lower upon acute exposure to a simulated altitude of 1,800 m/5,905 ft (77.8%) compared with sea level (90.6%). Furthermore, decrements in S_aO_2 have been demonstrated in endurance athletes at elevations as low as 580 m (1,900 ft). Gore et al. (1997) evaluated trained male and female cyclists and triathletes during a 5-min maximal cycle ergometer exercise test in a hypobaric chamber in conditions simulating sea level and low altitude (580 m/1,900 ft). Arterial oxyhemoglobin saturation decreased significantly in both the men (sea level = 92.0%, altitude = 90.1%) and women athletes (sea level = 92.1%, altitude = 89.7%) upon acute exposure to a simulated altitude of 580 m/1,900 ft (Gore et al. 1997). Similar findings were reported by Gore et al. (1996) in well-trained male cyclists, whose S_aO_2 fell significantly at a simulated altitude of 580 m/1,900 ft (86.5%) compared with sea level (90.4%). The effect of acute altitude exposure on arterial oxyhemoglobin saturation in endurance athletes is summarized in table 2.1 on pages 26-27.

Maximal Oxygen Consumption

It is well documented that maximal oxygen consumption is reduced upon exposure to altitude. In a classic study, Squires and Buskirk (1982) evaluated the effects of acute exposure to hypobaric hypoxia on $\dot{V}O_2$max in male recreational runners ($\dot{V}O_2$max = 60.1 ml · kg^{-1} · min^{-1}). Each runner performed a maximal treadmill test in a hypobaric chamber at simulated altitudes of 362 m (1,190 ft), 914 m (3,000 ft), 1,219 m (4,000 ft), 1,524 m (5,000 ft), and 2,286 m (7,500 ft). Maximal oxygen uptake at 362 m averaged 4.35 L · min^{-1} and was significantly reduced by 5%, 7%, and 12% at 1,219 m, 1,524 m, and 2,286 m, respectively (Squires and Buskirk 1982). Figure 2.2 illustrates the effect of acute hypobaric hypoxia on $\dot{V}O_2$max as reported by Squires and Buskirk (1982). Several additional studies involving athletes and physically fit soldiers have demonstrated that $\dot{V}O_2$max declines in a curvilinear manner as altitude increases from 580 m to 8,848 m (1,900-29,020 ft). These data have been summarized in a review article by Fulco et al. (1998) and are illustrated in figure 2.3 on page 28. Altitude-induced decrements in maximal oxygen consumption appear to be gender independent (Fulco et al. 1998).

Several studies have examined the effect of altitude exposure on $\dot{V}O_2$max in international-caliber endurance athletes at moderate elevations ranging from 1,500 m to 1,822 m (4,920-5,975 ft). Bailey et al. (1998) reported that among British male elite distance runners, maximal oxygen

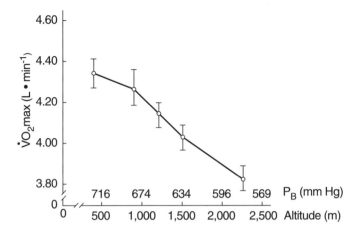

Figure 2.2 The effect of acute simulated altitude exposure from 362 m (1,190 ft) to 2,286 m (7,500 ft) on maximal oxygen consumption ($\dot{V}O_2$max). P_B = barometric pressure.

Reprinted, by permission, from R.W. Squires and E.R. Buskirk, 1982, "Aerobic capacity during acute exposure to simulated altitude, 914 to 2286 meters," *Medicine and Science in Sports and Exercise* 14: 38.

consumption measured during a treadmill test on the 19th day of training at 1,640 m (5,380 ft) was significantly reduced by 13% (5.15 vs. 4.48 L · min⁻¹) compared with prealtitude values. In a study of Australian national team male cross-country skiers (Roberts et al. 1998), $\dot{V}O_2$max measured during a multistage treadmill ski test at sea level (70.2 ml · kg⁻¹ · min⁻¹) was significantly lower by 12% upon acute exposure to a simulated altitude of 1,800 m/5,905 ft (61.7 ml · kg⁻¹ · min⁻¹). Among male elite rowers, $\dot{V}O_2$max was 15% below sea level values ($p < 0.05$) during the initial days of a 21-day altitude training camp at 1,820 m/5,970 ft (Jensen et al. 1993). Saltin et al. (1995) reported that maximal oxygen consumption declined by 15% (statistics not reported, n = 6) in Scandinavian male elite distance runners when measured during a maximal treadmill test at 2,000 m/6,560 ft (67.3 ml · kg⁻¹ · min⁻¹) compared with sea level (79.2 ml · kg⁻¹ · min⁻¹). The same pattern was observed among Kenyan male elite distance runners whose treadmill $\dot{V}O_2$max was reduced by 17% (statistics not reported, n = 6) at 2,000 m (66.3 ml · kg⁻¹ · min⁻¹) versus sea level (79.9 ml · kg⁻¹ · min⁻¹) (Saltin et al. 1995).

In studies conducted at 2,300 m (7,544 ft), Daniels and Oldridge (1970) reported that treadmill $\dot{V}O_2$max was reduced by 15% (statistics not reported, n = 6) in U.S. male elite distance runners after two days at altitude (sea level = 74.8, altitude = 63.6 ml · kg⁻¹ · min⁻¹), while Adams et al. (1975) indicated that maximal oxygen consumption was 17% lower

Table 2.1 Effect of Acute Altitude Exposure on Arterial Oxyhemoglobin Saturation

Author	Altitude (m/ft)	Subjects	Method	Exercise	Sea level oxyhemoglobin saturation (%)	Altitude oxyhemoglobin saturation (%)	Oxyhemoglobin saturation decrement (relative %)
Gore et al. 1996	580/1,900[a]	Male trained cyclists (n = 9)	S_aO_2	Cycle ergometer $\dot{V}O_2$max	90.4	86.5	4*
Gore et al. 1997	580/1,900[a]	Female trained cyclists (n = 10) Male trained cyclists (n = 10)	S_aO_2	Cycle ergometer $\dot{V}O_2$max	92.1 92.0	89.7 90.1	3* 2*
Squires and Buskirk 1982	915/3,000[a] 1,220/4,000 1,525/5,000	Male trained runners (n = 12)	S_pO_2	Treadmill $\dot{V}O_2$max	90.7	87.2	4*
Roberts et al. 1998	1,800/ 5,905[b]	Male elite CC skiers (n = 9) Australian NT	S_aO_2	Treadmill CC ski $\dot{V}O_2$max	90.6	77.8	14*

Brosnan et al. 2000	2,100/ 6,890[b]	Female elite cyclists (n = 8) Australian NT	S_pO_2	3 × 10-min cycling bouts (~85% $\dot{V}O_2$peak)	95.4	92.9	3*
Squires and Buskirk 1982	2,286/ 7,500[a]	Male trained runners (n = 12)	S_pO_2	Treadmill $\dot{V}O_2$max	90.7	79.1	13*
Lawler et al. 1988	3,000/ 9,840[a]	Trained endurance athletes (n = 7)	S_pO_2	Cycle ergometer $\dot{V}O_2$max	90.1	77.3	14*

Studies are listed by ascending altitude.

[a]Hypobaric chamber.
[b]Hypoxic gas mixture.
*Significant decrement ($p < 0.05$) in arterial oxyhemoglobin saturation at altitude versus sea level.

CC = cross-country; Hb = hemoglobin; NT = national team; S_aO_2 = oxyhemoglobin saturation measured via arterial blood samples; S_pO_2 = oxyhemoglobin saturation measured via pulse oximetry; $\dot{V}O_2$max = maximal oxygen consumption.

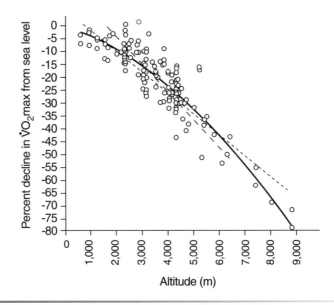

Figure 2.3 Percent $\dot{V}O_2$max decline with increasing elevation. Each of the 146 points (unfilled circles) on the graph represents a mean value derived from 67 different civilian and military investigations conducted at altitudes from 580 m (1,900 ft) to 8,848 m (29,020 ft). Multiple mean data points for a study were included if the study entailed more than one of the following: (a) elevation, (b) group of test subjects, or (c) exposure duration. For studies using hypoxic gas mixtures, inspired oxygen values were converted to altitude equivalents. Mean values are reported because they were the only values common to all investigations. A database regression line (the thick curvilinear line) was drawn using the 146 points. Since each of these data points is a mean value of many intra-investigation individual determinations of $\dot{V}O_2$max, the regression line represents possibly thousands of $\dot{V}O_2$max test values and therefore provides a true approximation for the expected average decrement at each elevation. Also included are the regression lines of Buskirk et al. (1967b) (dashed line) and Grover et al. (1986) (dotted line) that represent two of the most often quoted relationships between the decline in $\dot{V}O_2$max and an increase in elevation.

Reprinted, by permission, from C.S. Fulco et al., 1998, "Maximal and submaximal exercise performance at altitude," *Aviation, Space, and Environmental Medicine* 69: 794.

($p < 0.05$) versus sea level values in well-trained male distance runners after three days at altitude. At the same elevation, treadmill $\dot{V}O_2$max was 17% lower (statistics not reported, n = 5) in male collegiate distance runners after seven days at altitude (Faulkner et al. 1968), whereas maximal oxygen consumption measured during a tethered swimming test was 8% lower ($p < 0.05$) compared with sea level values in male collegiate swimmers after 14 days at altitude (Faulkner et al. 1967).

Several authors have investigated the effect of acute altitude exposure on maximal oxygen consumption at elevations exceeding 3,000 m (9,840 ft). Dill and Adams (1971) reported that treadmill $\dot{V}O_2$max was reduced by 18% (statistics not reported, n = 6) versus sea level values in well-trained male distance runners after two days at 3,090 m (10,135 ft). Similarly, Faulkner et al. (1968) demonstrated a 20% decrement (statistics not reported, n = 5) in male collegiate runners evaluated after five to seven days at 3,100 m (10,170 ft). At 4,000 m (13,120 ft), maximal oxygen consumption measured during a cycle ergometer test was 29% lower (statistics not reported, n = 6) compared with sea level values in well-trained male distance runners on the third day of altitude exposure (Buskirk et al. 1967a). Upon acute exposure to a simulated altitude of 4,000 m (13,120 ft), cycle ergometer $\dot{V}O_2$max was 23% lower (p < 0.05) in female and male elite triathletes compared with prealtitude values (Vallier et al. 1996). Faulkner et al. (1968) reported that treadmill $\dot{V}O_2$max was reduced by 29% (statistics not reported, n = 4) versus sea level values in male collegiate runners after three days at 4,300 m (14,110 ft).

It has been suggested that $\dot{V}O_2$max is reduced by approximately 9% for every 1,000-m (3,280-ft) increase in elevation above 1,050 m/3,445 ft (Robergs et al. 1998). However, recent evidence suggests that among trained endurance athletes, decrements in $\dot{V}O_2$max may be experienced at relatively low elevations. Peak oxygen consumption measured during a 5-min maximal cycle ergometer test decreased significantly in both male (5.21 vs. 4.90 L · min^{-1}; –6%) and female endurance athletes (3.56 vs. 3.43 L · min^{-1}; –4%) upon acute exposure to a simulated altitude of 580 m/1,900 ft (Gore et al. 1997). Similar findings were reported in well-trained male cyclists whose $\dot{V}O_2$max dropped 7% (p < 0.05) at a simulated altitude of 580 m (1,900 ft) (5.10 L · min^{-1}) compared with sea level (5.48 L · min^{-1}) (Gore et al. 1996).

A mathematical model has been proposed that describes the potential effect of acute altitude exposure on $\dot{V}O_2$max. According to the model of Peronnet et al. (1991), an athlete's maximal oxygen consumption at altitude, expressed as a percentage of sea level $\dot{V}O_2$max (% SL $\dot{V}O_2$max) at a given barometric pressure (P_B, expressed in torr), can be approximated using the following quadratic equation:

$$\% \text{ SL } \dot{V}O_2\text{max} = a_0 + a_1 P_B + a_2 (P_B^2) + a_3 (P_B^3)$$

where $a_0 = -174.1448622$, $a_1 = 1.0899959$, $a_2 = -1.5119 \times 10^{-3}$, and $a_3 = 0.72674 \times 10^{-6}$. This equation was developed from data obtained between 0 m and 4,000 m/13,120 ft (760-462 torr). Using the Peronnet et al. (1991) equation, for example, at an altitude of 2,000 m/6,560 ft (P_B ~600 torr), an athlete's maximal oxygen consumption would be approximately 93% of sea level $\dot{V}O_2$max. At an altitude of 3,000 m/9,840 ft (P_B ~525 torr), $\dot{V}O_2$max would be approximately 86% of the sea level value, whereas at 4,000 m/13,120 ft (P_B ~462 torr), $\dot{V}O_2$max would be approximately 78% of the sea level maximal oxygen consumption.

It should also be noted that there is great individual variability in the decrement in $\dot{V}O_2$max among athletes exposed to altitude, particularly among elite level athletes (Schouweiler and Stray-Gundersen 2002). In an effort to explain the individual variability in $\dot{V}O_2$max decrement, Koistinen et al. (1995) reported that the decrease in maximal oxygen consumption in trained athletes at a simulated altitude of 3,000 m (9,840 ft) was significantly correlated ($r = 0.61$) to sea level $\dot{V}O_2$max. In other words, the largest reduction in $\dot{V}O_2$max at altitude was observed in the most aerobically fit athletes. Several additional studies involving athletes and soldiers have supported the finding that relatively fit individuals ($\dot{V}O_2$max > 63 ml · kg^{-1} · min^{-1}) generally experience a greater decrement in $\dot{V}O_2$max at altitude than less fit individuals ($\dot{V}O_2$max < 51 ml · kg^{-1} · min^{-1}). These data have been summarized in a comprehensive review by Fulco et al. (1998) and are illustrated in figure 2.4. Billat et al. (2003)

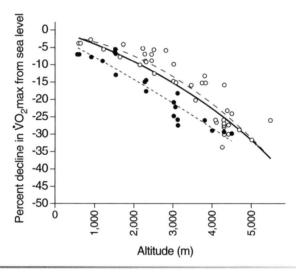

Figure 2.4 Effect of fitness level on $\dot{V}O_2$max decrement variability. This figure depicts results of studies of highly conditioned (baseline altitude $\dot{V}O_2$max ≥ 63 ml · kg^{-1} · min^{-1}; filled circles) and less conditioned ($\dot{V}O_2$max ≤ 51 ml · kg^{-1} · min^{-1}; unfilled circles) individuals. Only mean data points based on objective fitness criteria, that is, $\dot{V}O_2$max normalized to body weight and measured at the pre-exposure, resident altitude, are included. Studies using only descriptive terms such as "highly fit," "well-conditioned," or "trained" to characterize subjects were not included. To minimize possible confounding effects of altitude acclimatization or physical conditioning changes due to training while at altitude, data collected beyond the first three days of altitude exposure were also excluded. Regression lines for the highly conditioned (dotted line) and less conditioned (dashed line) individuals, as well as the database regression line (solid line) redrawn from figure 2.3 (but truncated at 5,500 m/18,045 ft), are included.

Reprinted, by permission, from C.S. Fulco et al., 1998, "Maximal and submaximal exercise performance at altitude," *Aviation, Space, and Environmental Medicine* 69: 794.

recently reported that among trained distance runners at a simulated altitude of 2,400 m (7,870 ft), there was an inverse relationship between running velocity at $\dot{V}O_2$max ($v\dot{V}O_2$max) and total time to exhaustion at $v\dot{V}O_2$max. In other words, the runners who showed the greatest altitude-induced decrement in $v\dot{V}O_2$max were also able to run the longest at $v\dot{V}O_2$max at altitude. The authors concluded that the longer run time to exhaustion by these runners who had the greatest drop in $v\dot{V}O_2$max was caused by a significant anaerobic energy contribution that was unaffected by altitude (Billat et al. 2003).

The greater decrement in $\dot{V}O_2$max seen in aerobically fit versus unfit individuals at altitude has been associated with pulmonary gas exchange limitations, which in turn result in a decrease in S_aO_2 and a reduction in oxygen availability to the exercising muscles. Lawler et al. (1988) reported that when sea level and simulated altitude conditions (3,000 m/9,840 ft) were compared, trained subjects experienced significantly greater decrements versus untrained subjects in relative $\dot{V}O_2$max (trained = –13.4, untrained = –4.6 ml · kg^{-1} · min^{-1}) and peak power output (trained = –50, untrained = –6 W). Similarly to what occurred in other studies, a significant positive correlation (r = 0.94) was found between sea level $\dot{V}O_2$max and sea level-to-altitude $\dot{V}O_2$max decrement. Furthermore, a significant negative correlation (r = –0.84) was reported for altitude S_pO_2 and sea level-to-altitude $\dot{V}O_2$max decrement. Collectively, the data of Lawler et al. (1988) suggested that trained aerobic athletes experience greater impairment of arterial oxyhemoglobin saturation and $\dot{V}O_2$max upon exposure to altitude than untrained individuals.

Results from a recent study by Chapman et al. (1999) suggested that the observed decrement in $\dot{V}O_2$max seen in well-trained endurance athletes training at altitude may be related more to the degree of exercise-induced arterial hypoxemia (EIAH) measured at sea level than to the athletes' level of fitness. Causes of EIAH include an attenuation of hyperventilation or an excessively wide alveolar-arterial oxygen pressure difference (A-aDO_2) (Dempsey and Wagner 1999; Derchak et al. 2000; Wetter et al. 2002). Several mechanisms have been proposed for the A-aDO_2 associated with EIAH: (1) ventilation-perfusion (V_A/Q) mismatch caused by the development of mild interstitial pulmonary edema or inflammatory-induced vascular leakage (Anselme et al. 1994; Prefaut et al. 1997; Sheel et al. 2001); (2) diffusion limitation caused by a reduction in alveolar capillary surface area, distance, or transit time caused by training-induced increases in pulmonary blood flow or possibly the development of mild interstitial pulmonary edema (Dempsey and Wagner 1999; Sheel et al. 2001; Wetter et al. 2002); and (3) veno-arterial shunts (Dempsey and Wagner 1999; Wetter et al. 2002). In the investigation by Chapman et al. (1999), well-trained male endurance athletes were divided into two groups based on their S_pO_2 response during maximal treadmill exercise at sea level. One group (n = 8, $\dot{V}O_2$max = 71.1 ml · kg^{-1} · min^{-1}) was classified as exercise-induced arterial hypoxemic (EIAH) based on S_pO_2 < 90% at $\dot{V}O_2$max measured at sea level. The other group (n = 6, $\dot{V}O_2$max = 67.2 ml · kg^{-1} · min^{-1}) was classified as non-exercise-induced arterial hypoxemic (non-EIAH) based on S_pO_2 > 92% at $\dot{V}O_2$max measured at sea level. Both groups completed an exhaustive incremental treadmill test in a simulated altitude environment (normobaric hypoxia, F_IO_2 = 0.1870) equivalent to 1,000 m (3,280 ft). Although the two groups were equally well trained, the EIAH runners demonstrated a significant 4% decline in $\dot{V}O_2$max from sea level to altitude, whereas the non-EIAH runners did not show a significant drop in $\dot{V}O_2$max. Thus, these data suggested that the degree of EIAH experienced by endurance athletes at sea level may be a more important factor than fitness level in contributing to the sea level-to-altitude decrement in $\dot{V}O_2$max. It is possible that the same physiological mechanisms that differentiate the non-EIAH athlete from the EIAH athlete at sea level also help to attenuate decrements in $\dot{V}O_2$max in the non-EIAH athlete at altitude. Whether these protective mechanisms occur in non-EIAH endurance athletes at elevations higher than 1,000 m (3,280 ft) is unclear.

Robergs et al. (1998) proposed a multivariate model to explain the individual variability in $\dot{V}O_2$max decrement at altitude. This model suggests that the greatest reduction in $\dot{V}O_2$max upon exposure to acute hypobaric hypoxia occurs in individuals who have a large sea level $\dot{V}O_2$max and (1) possess a low sea level lactate threshold, (2) experience a large reduction in S_aO_2 during maximal exercise at altitude, (3) have a relatively large lean body mass, and (4) are male. Less pronounced decrements in altitude $\dot{V}O_2$max are observed in individuals who have a large sea level $\dot{V}O_2$max but (1) possess a high sea level lactate threshold, (2) do not experience a large reduction in S_aO_2 during maximal exercise at altitude, and (3) have a relatively small lean body mass (Robergs et al. 1998).

In summary, data from several scientific studies suggest that maximal oxygen consumption is significantly reduced upon exposure to altitude. The subjects who participated in the investigations reviewed in this section were either well-trained or elite aerobic athletes. Most were evaluated for changes in $\dot{V}O_2$max within the initial days of exposure to actual or simulated elevations ranging from 580 m to 4,300 m (1,900-14,110 ft), whereas a few were tested after 7 to 20 days at altitude. All of these studies reported decrements in $\dot{V}O_2$max ranging from 2% to 29% depending on the altitude. It has been suggested that $\dot{V}O_2$max is reduced by approximately 9% for every 1,000-m (3,280-ft) increase in altitude above 1,050 m (3,445 ft). However, recent evidence suggests that among trained endurance athletes, decrements in $\dot{V}O_2$max may be experienced at elevations as low as 580 m (1,900 ft). In addition, it is highly likely that there is great individual variability in the decrement in $\dot{V}O_2$max among athletes exposed to altitude. Decrements in maximal oxygen consumption at altitude may be due in part to a hypoxia-induced reduction in arterial oxyhemoglobin saturation secondary to pulmonary capillary transit time limitations. Table 2.2 on pages 34-35 summarizes the findings of scientific studies that have evaluated the effect of altitude exposure on $\dot{V}O_2$max in trained aerobic athletes.

Aerobic Performance

Decrements in aerobic performance at altitude have been associated with altitude-induced reductions in $\dot{V}O_2$max secondary to reductions in S_aO_2 (figure 2.5, p. 36). Gore et al. (1997) demonstrated that significant decrements in aerobic performance may occur at elevations as low as 580 m (1,900 ft) in well-trained male and female cyclists and triathletes. These athletes performed a 5-min maximal work test on a cycle ergometer in a hypobaric chamber in environmental conditions simulating sea level and low altitude. Compared with values at sea level, total work was significantly reduced at 580 m (1,900 ft) by 4% in both the men (sea level = 115, altitude = 110 kilojoules [kJ]) and women athletes (sea level = 81, altitude = 78 kJ).

Similar results have been reported in recent studies examining athletic performance in elite endurance athletes at moderate altitudes ranging from 1,500 m to 1,822 m (4,920-5,975 ft). Bailey et al. (1998) evaluated British male elite distance runners and observed that running velocity during a 1,000-m time trial was significantly slower by 3% (sea level = 5.67, altitude = 5.52 m · sec^{-1}) and 4% (sea level = 5.79, altitude = 5.55 m · sec^{-1}) after approximately 14 days of training at 1,500 m (4,920 ft) and 1,640 m (5,380 ft), respectively. Australian national team male cross-country skiers demonstrated a 7% decrease (no statistically significant difference; NSD) relative to sea level values in the performance of a 45-min cross-country ski test upon acute exposure to a simulated altitude of 1,800 m/5,905 ft (Roberts et al. 1998). Among male elite rowers, work capacity (kJ) during a maximal rowing ergometer test was 7% below sea level values (p < 0.05) during the initial days of a 21-day altitude training camp at 1,822 m/5,975 ft (Jensen et al. 1993).

Additional investigations have been conducted at elevations exceeding 2,000 m (6,560 ft). Brosnan et al. (2000) recently reported that average power output was significantly lower at a simulated altitude of 2,100 m (6,890 ft) versus sea level in Australian elite female road cyclists during

Table 2.2 Effect of Altitude Exposure on $\dot{V}O_2$max in Trained Endurance Athletes

Author	Altitude (m/ft)	Subjects	Altitude exposure prior to $\dot{V}O_2$max test (days)	$\dot{V}O_2$max decrement (%)
Gore et al. 1996	580/1,900[a]	Male trained cyclists (n = 9)	a	7*
Gore et al. 1997	580/1,900[a]	Female trained cyclists (n = 10) Male trained cyclists (n = 10)	a a	4* 6*
Squires and Buskirk 1982	914/3,000[a]	Male trained runners (n = 12)	a	2 NSD
Chapman et al. 1999	1,000/3,280[b]	EIAH male trained runners (n = 8) Non-EIAH male trained runners (n = 6)	b	4* 1 NSD
Squires and Buskirk 1982	1,219/4,000[a] 1,524/5,000[a]	Male trained runners (n = 12)	a a	5* 7*
Bailey et al. 1998	1,640/5,380	Male elite runners (n = 10) British NT	19-20	13*
Roberts et al. 1998	1,800/5,905[b]	Male elite CC skiers (n = 9) Australian NT	a	12*
Jensen et al. 1993	1,822/5,980	Male elite rowers (n = 9) Italian NT	≤5	15*
Saltin et al. 1995	2,000/6,560	Male Scandinavian elite runners (n = 6) Male Kenyan elite runners (n = 6)	1-5 c	15* 17*
Squires and Buskirk 1982	2,286/7,500[a]	Male trained runners (n = 12)	a	12*

Study	Altitude	Subjects		
Faulkner et al. 1967	2,300/7,544	Male collegiate swimmers (n = 15)	14	8*
Faulkner et al. 1968	2,300/7,544	Male collegiate runners (n = 5)	7	17#
Daniels and Oldridge 1970	2,300/7,544	Male elite runners (n = 6) U.S. NT	2	15*
Adams et al. 1975	2,300/7,544	Male trained runners (n = 12)	3	17*
Dill and Adams 1971	3,090/10,135	Male trained runners (n = 6)	2	18#
Faulkner et al. 1968	3,100/10,170	Male collegiate runners (n = 5)	5-7	20*
Buskirk et al. 1967a	4,000/13,120	Male collegiate runners (n = 6)	3	29*
Vallier et al. 1996	4,000/13,120[a]	Female and male elite triathletes (n = 5) French NT	1	28*
Faulkner et al. 1968	4,300/14,110	Male collegiate runners (n = 4)	3	29*

Studies are listed by ascending altitude.

[a]Hypobaric chamber.
[b]Hypoxic gas mixture.
[c]Altitude residents.
*Significant decrement (p < 0.05) in $\dot{V}O_2$max at altitude versus sea level.
#HyStatistical analysis not reported.

CC = cross-country; EIAH = exercise-induced arterial hypoxemic (S_pO_2 < 90% measured at sea level $\dot{V}O_2$max); NT = national team; NSD = no statistically significant difference; $\dot{V}O_2$max = maximal oxygen consumption.

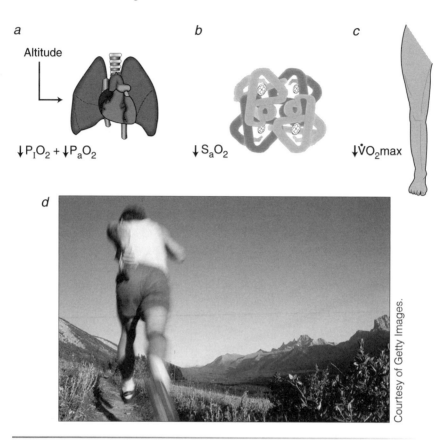

Figure 2.5 Sequence of physiological responses that occur at altitude leading to reductions in aerobic performance and training capacity: *(a)* altitude-induced decrements in the partial pressure of inspired oxygen (P_IO_2) and arterial oxygen (P_aO_2), *(b)* decrease in arterial oxyhemoglobin saturation (S_aO_2), *(c)* decrease in maximal oxygen consumption ($\dot{V}O_2max$), and *(d)* decrease in aerobic performance and training capacity.

Parts *(b)* and *(c):* Fig. 18.4, p. 582 from *Human Anat my and P ysi l gy,* 2nd ed. by Elaine N. Marieb. Copyright © 1992 by the Benjamin/Cummings Publishing Company, Inc. Reprinted by permission of Pearson Education, Inc.

each of three 10-min endurance cycling bouts done at approximately 85% $\dot{V}O_2$peak ([1] 226 vs. 244 W; [2] 221 vs. 234 W; [3] 221 vs. 235 W). Many of the original studies on the topic of altitude training were conducted at 2,300 m (7,544 ft) in an attempt to simulate the environmental conditions of the 1968 Mexico City Olympics (2,300 m/7,544 ft). Faulkner et al. (1967) evaluated the effect of acute altitude exposure among male collegiate swimmers and reported that swim performance in 100-yd, 200-yd, and 500-yd time trials was significantly slower versus sea level performance by 2%, 5%, and 6%, respectively, after one day at 2,300 m

(7,544 ft). At the same elevation, Daniels and Oldridge (1970) reported that 1-mile and 3-mile run times were slower (statistics not reported, n = 6) by 20 sec (sea level = 4:06, altitude = 4:26) and 83 sec (sea level = 14:02, altitude = 15:25), respectively, in U.S. male elite distance runners after two days at altitude, whereas Adams et al. (1975) indicated that 2-mile run time was 7% slower (statistics not reported, n = 12) versus that at sea level in well-trained male distance runners after three days at altitude. After seven days at 2,300 m (7,544 ft), time trial performance (1-3 miles) was 7% slower (statistics not reported, n = 5) in male collegiate distance runners (Faulkner et al. 1968).

Several authors have investigated the effect of acute altitude exposure on endurance performance at elevations exceeding 3,000 m (9,840 ft). Dill and Adams (1971) reported that treadmill run time to exhaustion was slower by 27% (sea level = 9:20, altitude = 6:49) compared with that at sea level in well-trained male distance runners at 3,090 m/10,135 ft (statistics not reported, n = 6). At 4,000 m (13,120 ft), total ride time in a maximal bicycle ergometer test by well-trained male distance runners was 12% less (statistics not reported, n = 6) versus that at sea level on the third day of altitude exposure (Buskirk et al. 1967a). In addition, Buskirk et al. (1967a) indicated that all the runners were slower versus their prealtitude times over several middle and long distances including 440 yd (9%), 880 yd (18%), 1 mile (23%), and 2 miles (19%) after approximately six weeks of training at 4,000 m (13,120 ft). Vallier et al. (1996) reported that power output during an incremental exhaustive cycle ergometer test was 23% lower ($p < 0.05$) compared with values at sea level in female and male elite triathletes upon acute exposure to a simulated altitude of 4,000 m (13,120 ft). Competitive marathon performance by elite male distance runners has been shown to be significantly slower at the relatively high elevations of 4,300 m/14,110 ft (12.3 km · hr^{-1}, 3 hr:25 min:48 sec, –29%) and 5,200 m/17,055 ft (11.5 km · hr^{-1}, 3 hr:40 min:12 sec, –34%) compared with sea level marathon performance (17.3 km · hr^{-1}, 2 hr:26 min:24 sec) after seven days (4,300 m) and three weeks (5,200 m) at altitude (Roi et al. 1999). World-best marathon performances at elevations ranging from 0 m (Rotterdam, Netherlands) to 5,200 m/17,055 ft (Tibet) are listed in table 2.3 (Roi et al. 1999).

As previously described, a mathematical model has been proposed that quantifies the potential effect of altitude on $\dot{V}O_2$max (Peronnet et al. 1991). This model can also be used to calculate the effect of altitude on running performance. It suggests that in running events exceeding 400 m, there is an exponential deterioration in performance due to the altitude-induced reduction in $\dot{V}O_2$max. This decrement in running performance is more pronounced in long distance events (5,000 m, 10,000 m, marathon) versus middle distance events (800 m, 1,500 m) due to an increased reliance on aerobic metabolism in the longer

Table 2.3 Best Marathon Performances by Men at Different Altitudes

Year	Location	Altitude (m/ft)	Time (hr:min:sec)	% Sea level time	Winner
1998	Rotterdam	0/0	2:06:50	100	Dinsamo (Ethiopia)
1993	Mexico City	2,240/ 7,345	2:14:47	94	Ceron (Mexico)
1994	Tibet	4,300/ 14,110	2:56:08	72	Carpenter (USA)
1995	Tibet	5,200/ 17,055	3:22:25	62	Carpenter (USA)

Adapted, by permission, from G.S. Roi et al., 1999, "Marathons at altitude," *Medicine and Science in Sports and Exercise* 31: 726.

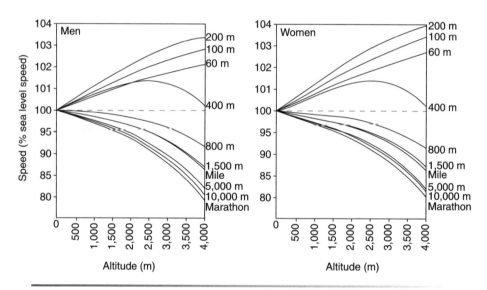

Figure 2.6 The effect of altitude (up to 4,000 m/13,120 ft) on running velocity in events ranging from 100 m to the marathon in men (left panel) and women (right panel).

Reprinted, by permission, from F. Peronnet et al., 1991, "A theoretical analysis of the effect of altitude on running performance," *Journal of Applied Physiology* 70: 402.

events. For example, in Mexico City (2,300 m/7,544 ft), the decrement in average running velocity is approximately 2% for 800 m, 4% for 1,500 m, 6% for 5,000 m, and 7% for the marathon, for both men and women (Peronnet et al. 1991). Figure 2.6 shows the effect of acute altitude exposure on running velocity over various distances for both men and women. Performance results from the 1968 Mexico City Summer Olympics tend to support this model. In men's track and field, world records were established in the long jump and most of the sprint events. In contrast, the performances of gold medalists in the long distance events (>1,500 m) were substantially slower than the 1968 world records. Similar performance results were seen in women's track and field at the Mexico City Olympics.

In sports in which *aerodynamics* play an important role, performance may be enhanced when competition takes place at altitude as a consequence of reduced air density and decreased aerodynamic resistance (Hahn and Gore 2001). It has been suggested that the optimal altitude for cycling time trials of 2-km to 40-km length is 3,200 m to 3,500 m (10,495-11,480 ft), with times estimated to be 4% to 4.5% faster compared with times for sea level performances (Olds 1992, 2001). A model proposed by Capelli and di Prampero (1995) suggests that the optimal elevation for breaking the world record for 1.0-hr unaccompanied cycling is approximately 4,000 m (13,120 ft). At this elevation there is a significant reduction in $\dot{V}O_2$max. However, there is a greater reduction in the amount of power that is dissipated by the cyclist against air resistance. In other words, in cycling competition at altitude, the aerodynamic drag is less because of the decrease in air density (figure 2.7). This improvement in cycling aerodynamics exceeds the decrement in $\dot{V}O_2$max and results in greater cycling velocity at altitude versus sea level. For the 1.0-hr cycling time trial, the model proposed by Capelli and di Prampero (1995) suggests that 4,000 m (13,120 ft) is the altitude at which an optimal aerodynamic-$\dot{V}O_2$max trade-off is achieved. Beyond this elevation, the decrement in maximal oxygen consumption may exceed aerodynamic gains and thus lead to a reduction in cycling velocity. More recently, Bassett and colleagues (1999) have proposed a mathematical model that suggests the optimal elevation for setting the 1.0-hr world record is approximately 2,500 m (8,200 ft) for acclimatized cyclists and about 2,000 m (6,560 ft) for unacclimatized cyclists (figure 2.8). This is roughly the altitude of Mexico City where the 1.0-hr world record was set by Ole Ritter (1968), Eddy Merckx (1972), and Francesco Moser (twice in 1984) (Kyle and Bassett 2003). The highest velodrome in the world is located in La Paz, Bolivia, at 3,417 m (11,210 ft). Interestingly, the current 1.0-hr world record (56.375 km), held by Chris Boardman of Great Britain, was set on an indoor track near sea level in Manchester, England (Hahn and Gore 2001).

During the 2002 Winter Olympics, the long-track speedskating events were contested at the Utah Olympic Oval, located in the Kearns section of Salt Lake City at 1,425 m (4,675 ft). In addition to providing aerodynamic

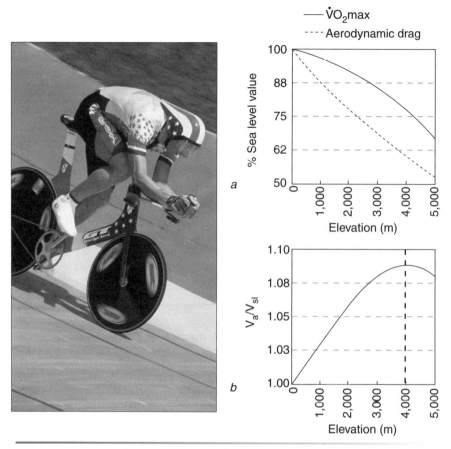

Figure 2.7 The effect of altitude on cycling performance in the 1-hr time trial. (a) Although $\dot{V}O_2$max (solid line) decreases as one ascends from sea level to 5,000 m (16,400 ft), the amount of aerodynamic drag (broken line) is reduced to a greater degree, allowing the cyclist to ride faster at altitude despite the drop in $\dot{V}O_2$max. (b) The "$\dot{V}O_2$max decrement versus aerodynamic benefit" is optimized at an elevation of approximately 4,000 m (13,120 ft), after which point the decrement in $\dot{V}O_2$max exceeds the beneficial effects of reduced aerodynamic drag. V_a/V_{sl} = velocity at a specific altitude relative to velocity at sea level; $\dot{V}O_2$max = maximal oxygen consumption.

Reprinted, by permission, from C. Capelli and P.E. di Prampero, 1995, "Effects of altitude on top speeds during 1 h unaccompanied cycling," *European Journal of Applied Physiology* 71: 470.

benefits, the altitude of Salt Lake City produced changes in the ice surface that further enhanced the speedskaters' performance. At 1,425 m, the barometric pressure is approximately 84% of that at sea level. This means that about 16% less air will dissolve in water. Because there was less air dissolved in the water, there were fewer air bubbles in the ice that covered the Utah Olympic Oval, which resulted in a smoother,

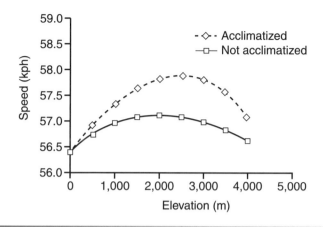

Figure 2.8 Theoretical effect of altitude on the hour record for acclimatized and unacclimatized cyclists.
Reprinted, by permission, from C.R. Kyle and D.R. Bassett, 2003, The cycling world hour record. In *High-tech cycling*, 2nd ed., edited by E.R. Burke (Champaign, IL: Human Kinetics), 195.

harder, and faster surface versus that at similar speedskating venues at sea level. In addition, at altitude it takes less skate pressure to melt ice under the blade and thereby creates a small pool of water that allows the skater to hydroplane and skate faster.

The Utah Olympic Oval established itself as the "fastest ice in the world" during the 2001 World Championships when five world records were set in events ranging from the 500 m to 5,000 m. Performances at the 2002 Salt Lake City Winter Olympics were even more impressive. In the 10 events ranging from 500 m to 10,000 m, a total of eight world records were established. Table 2.4 on page 42 summarizes the results of long-track speedskating at the 2002 Winter Olympics.

In summary, scientific data as well as mathematical models suggest that aerobic performance is negatively affected upon exposure to altitude. In general, the subjects recruited for the scientific studies reviewed in this section were either well-trained or elite endurance athletes who were evaluated on aerobic performance within the initial days of exposure to actual or simulated elevations ranging from 580 m to 5,200 m (1,900-17,055 ft). Performance measures in these studies included sport-specific time trials, work capacity tests conducted on sport-specific ergometers, and endurance time during an exhaustive incremental exercise test. All of these investigations showed decrements in aerobic performance upon acute exposure to altitude. Performance decrements ranged from 2% to 34% depending on the altitude. Decrements in aerobic performance at altitude have been associated with hypoxia-induced reductions in $\dot{V}O_2$max secondary

Table 2.4 Results of the Long-Track Speedskating Events at the 2002 Salt Lake City Winter Olympics

Event	Olympic champion	Time (min:sec)	Previous or current world record (min:sec)	Difference (%)
W 500 m	LeMay Doan (Canada)	0:37.29	0:37.22	+0.2
M 500 m	Fitzrandolph (USA)	0:34.42	0:34.32	+0.3
W 1,000 m	Witty (USA)	1:13.83*	1:14.06	−0.3
M 1,000 m	van Velde (Netherlands)	1:07.18*	1:07.72	−0.8
W 1,500 m	Friesinger (Germany)	1:54.02*	1:54.38	−0.3
M 1,500 m	Parra (USA)	1:43.95*	1:45.20	−1.2
W 3,000 m	Pechstein (Germany)	3:57.70*	3:59.26	−0.7
W 5,000 m	Pechstein (Germany)	6:46.91*	6:52.44	−1.3
M 5,000 m	Uytdehaage (Netherlands)	6:14.66*	6:18.72	−1.1
M 10,000 m	Uytdehaage (Netherlands)	12:58.92*	13:03.40	−0.6

Events contested at the Utah Olympic Oval (1,425 m/4,675 ft).
*World and Olympic record.

to reductions in S_aO_2. In sports such as cycling and speedskating in which aerodynamics play an important role, the decrement in $\dot{V}O_2$max may be offset by aerodynamic advantages up to elevations of approximately 4,000 m (13,120 ft). Table 2.5 on pages 44-45 summarizes the results of several studies that have evaluated the effect of altitude exposure on endurance performance in trained aerobic athletes.

Training Capacity

In addition to decrements in aerobic performance, there is also a decrease in training capacity upon exposure to altitude. Because of altitude-induced decrements in S_aO_2 and $\dot{V}O_2$max, it is difficult for endurance athletes to maintain sea level training intensity at higher elevations. Although some athletes attempt to replicate the absolute training load of their sea level workouts, they do so at the risk of becoming ill, injured, or overtrained. Both anecdotal and scientific evidence suggests that it is difficult for most athletes to maintain sea level intensity during altitude training workouts, especially in the initial days at altitude.

Levine and Stray-Gundersen (1997) quantified the effect of altitude exposure on training intensity in competitive distance runners who were divided into three groups: (1) a "low-low" group that lived and trained at sea level (150 m/490 ft); (2) a "high-low" group that lived at moderate altitude (2,500 m/8,200 ft) and trained at low altitude (1,250 m/4,100 ft); and (3) a "high-high" group that lived and trained at moderate altitude (2,500 m/8,200 ft). Table 2.6 on page 46 shows the differences in training intensity among the three groups during base/over-distance training as well as interval training (1,000-m intervals). Similar results were recently reported by Niess et al. (2003) in an investigation of male and female distance runners ($\dot{V}O_2$max = 64.6 ml · kg^{-1} · min^{-1}) who performed a high-intensity interval training workout at sea level and following six days of acclimatization at 1,800 m (5,905 ft). The interval training session consisted of 10 × 1,000 m with a 2-min recovery, and it was performed with an equivalent blood lactate accumulation pattern at sea level and altitude. Despite the similar blood lactate levels during the interval workout at sea level and altitude, mean running velocity (m · sec^{-1}) and running velocity expressed as a percentage of $\dot{V}O_2$max (%$\dot{V}O_2$max) were significantly lower by 4% and 5%, respectively, at 1,800 m (5,905 ft) compared with sea level. Collectively, the results of these studies suggested that absolute training intensity during base and interval workouts is significantly reduced at low and moderate altitude in well-trained competitive distance runners.

It appears that some athletes may be able to maintain sea level training intensity during interval workouts conducted at moderate altitude. Using a retrospective analysis, Chapman et al. (1998) classified female and male collegiate runners from three altitude training groups: (1) "high-low" (lived at 2,500 m/8,200 ft; trained at 1,250 m/4,100 ft), (2) "high-high" (lived and trained at 2,500 m/8,200 ft), and (3) "high-high-low" (lived at 2,500 m/8,200 ft; base training at 2,500 m/8,200 ft; interval training at 1,250 m/4,100 ft) into responders or nonresponders based on their performance in a postaltitude 5-km run. Training data obtained on the 14th day at altitude (1,250 m/4,100 ft) showed that the responders' interval training velocity at altitude (5.0 m · sec^{-1}; 1,000-m intervals) was similar to their prealtitude sea level training velocity (5.1 m · sec^{-1}). In contrast, the nonresponders' training velocity at altitude (4.8 m · sec^{-1}) was significantly slower versus their prealtitude sea level training velocity (5.2 m · sec^{-1}). One should note that the responders'

Table 2.5 Effect of Altitude Exposure on Endurance Performance in Trained Aerobic Athletes

Author	Altitude (m/ft)	Subjects	Performance test	Altitude exposure prior to performance test (days)	Performance decrement (%)
Gore et al. 1997	580/ 1,900[a]	Female trained cyclists (n = 10)	5-min maximal cycling test (kJ)	[a]	4*
		Male trained cyclists (n = 10)	5-min maximal cycling test (kJ)	[a]	4*
Bailey et al. 1998	1,500/ 4,920	Male elite runners (n = 10) British NT	1,000-m TT (m · sec^{-1})	16	3*
	1,640/ 5,380	Male elite runners (n = 10) British NT	1,000-m TT (m · sec^{-1})	18	4*
Roberts et al. 1998	1,800/ 5,905[b]	Male elite CC skiers (n = 9) Australian NT	45-min CC ski test (m)	[a]	7 NSD
Jensen et al. 1993	1,822/ 5,975	Male elite rowers (n = 9) Italian NT	6-min maximal rowing test (kJ)	≤5	7*
Brosnan et al. 2000	2,100/ 6,890[b]	Female elite cyclists (n = 8) Australian NT	3 × 10-min cycling bouts at ~85% $\dot{V}O_2$peak (W)	[b]	6*
Faulkner et al. 1967	2,300/ 7,544	Male collegiate swimmers (n = 15)	100-yd TT (min:sec)	1	2*
			200-yd TT (min:sec)	1	5*
			500-yd TT (min:sec)	1	6*
Faulkner et al. 1968	2,300/ 7,544	Male collegiate runners (n = 5)	1-, 2-, and 3-mile TT (min:sec)	7	7#
Daniels and Oldridge 1970	2,300/ 7,544	Male elite runners (n = 6) U.S. NT	1-mile TT (min:sec)	2	8#
			3-mile TT (min:sec)	2	10#

Study	Altitude (ft)	Subjects	Test	Days at altitude	% decrement
Adams et al. 1975	2,300/ 7,544	Male trained runners (n = 12)	2-mile TT (min:sec)	3	7[#]
Dill and Adams 1971	3,090/ 10,135	Male trained runners (n = 6)	TM run time to exhaustion (min)	2-16	27[#]
Buskirk et al. 1967a	4,000/ 13,120	Male collegiate runners (n = 6)	Cycling time to exhaustion (min)	3	12[#]
			440-yd TT (min:sec)	40-57	9[#]
			880-yd TT (min:sec)	40-57	18[#]
			1-mile TT (min:sec)	40-57	23[#]
			2-mile TT (min:sec)	40-57	19[#]
Vallier et al. 1996	4,000/ 13,120[a]	Female and male elite triathletes (n = 5) French NT	Maximal cycling test (W)	1	28*
Roi et al. 1999	4,300/ 14,110	Male elite marathon runners (n = 5)	Competitive marathon (km · hr^{-1})	7	29*
Roi et al. 1999	5,200/ 17,055	Male elite marathon runners (n = 5)	Competitive marathon (km · hr^{-1})	21	34*

Studies are listed by ascending altitude.

[a] Hypobaric chamber.
[b] Hypoxic gas mixture.
* Significant decrement ($p < 0.05$) in endurance performance at altitude versus sea level.
[#] Statistical analysis not reported.

CC = cross-country; kJ = kilojoule; km = kilometer; NSD = no statistically significant difference; NT = national team; TM = treadmill; TT = time trial; W = watt.

Table 2.6 Effect of Altitude on Absolute Training Intensity in Competitive Distance Runners

	Base training		Interval training	
	Running velocity (% SL 5,000-m time)	$\dot{V}O_2$ (% SL $\dot{V}O_2$max)	Running velocity (% SL 5,000-m time)	$\dot{V}O_2$ (% SL $\dot{V}O_2$max)
Low-low (n = 13)	82	72	111	92
High-low (n = 13)	77*	67	104*	86
High-high (n = 13)	76*	64*	96*	74*

*Significantly different versus low-low (p < 0.05)

Low-low: lived and trained at sea level (150 m/490 ft); high-low: lived at moderate altitude (2,500 m/8,200 ft) and trained at low altitude (1,250 m/4,100 ft); high-high: lived and trained at moderate altitude (2,500 m/8,200 ft); SL = sea level; $\dot{V}O_2$max = maximal oxygen consumption.

Adapted, by permission, from B.D. Levine and J. Stray-Gundersen, 1997, "Living high-training low: Effect of moderate-altitude acclimatization with low-altitude training on performance," *Journal of Applied Physiology* 83: 107.

1,000-m velocity at altitude was similar to their sea level 1,000-m velocity when measured after 14 days of living and training at a minimum of 1,250 m (4,100 ft). It is possible, however, that the responders' 1,000-m velocity was significantly slower versus their sea level 1,000-m velocity during the initial week at altitude.

Recent studies suggest that altitude-induced decrements in training capacity can be offset through utilization of supplemental oxygen, which can simulate either normoxic (i.e., sea level) or hyperoxic conditions during high-intensity workouts conducted at altitude (Chick et al. 1993; Morris et al. 2000; Wilber et al. 2003). Supplemental oxygen training is described in more detail in chapter 6, "Current Practices and Trends in Altitude Training."

Anaerobic Performance

In contrast to the situation with aerobic performance, there are fewer scientific investigations of the effect of acute altitude exposure on anaerobic performance. Most of these studies evaluated either untrained or moderately trained individuals; only one investigation (Brosnan et al. 2000) measured the effect of acute hypoxia on anaerobic performance in elite athletes. These laboratory-controlled studies have used a variety of experimental altitudes and have defined anaerobic performance as either running sprint velocity (m · sec^{-1}) or power output (W or W · kg^{-1}) during maximal or supramaximal cycling exercise. At present, the results of these scientific studies are equivocal regarding the effect of acute altitude exposure on running sprint velocity and power output. The following discussion of investigations is ordered according to ascending altitude.

A few studies have evaluated the effect of moderate altitude (2,100 m/6,890 ft to 3,000 m/9,840 ft) on anaerobic performance. Brosnan et al. (2000) recently tested Australian elite female road cyclists at a simulated altitude of 2,100 m (6,890 ft) during the performance of three sets of 6 × 15-sec sprints (work:recovery ratios were 1:3, 1:2, and 1:1, respectively, for the three sets). Average power output (W) was similar in hypoxia versus normoxia in the first sprint set, but was significantly lower in the second (452 vs. 429 W) and third sprint set (403 vs. 373 W). Blood pH and HCO_3^- were significantly lower after each of the three sprint sets in hypoxia, suggesting a respiratory alkalosis. Balsom et al. (1994) evaluated trained male physical education students who completed 10 × 6-sec cycling sprints (30-sec recovery) at a simulated altitude of 3,000 m (9,840 ft). Not surprisingly, average power output (W) declined over the 10 sprints in both normoxia and hypoxia. However, average power output in sprints 8 and 9 was significantly lower in the hypoxic trial versus the normoxic trial. Blood lactate concentration was significantly higher in the hypoxic trial when measured after sprint 5 (6.9 vs. 5.8 mmol · L^{-1}) and sprint 10 (10.3 vs. 8.5 mmol · L^{-1}).

A limited number of studies have used hypoxic gas mixtures to evaluate the effect of relatively high altitude (4,270 m/14,005 ft to 5,790 m/18,990 ft) on anaerobic performance. Readers should note that the subjects in all of these studies were untrained individuals; therefore, extrapolation of the results to elite athletes is probably not valid. Weyand et al. (1999) tested four males at a simulated altitude of 4,270 m (14,005 ft) while they performed 15 maximal running sprints lasting from 15 to 180 sec. Recovery varied from 10 to 20 min depending on the length of the sprint. Sprint velocity in hypoxia was similar to that in normoxia in the shorter sprints of 15 sec (7.0 vs. 7.0 m · sec^{-1}) and 60 sec (5.4 vs. 5.5 m · sec^{-1}). However, sprint velocity was significantly slower in hypoxia versus normoxia in the 75-sec (0.2 m · sec^{-1} slower) and 180-sec sprints (0.7 m · sec^{-1} slower). McClellan et al. (1993) evaluated the performance of a 45-sec Wingate test (resistance = 7.5% total body weight) by 12 men at a simulated altitude of 5,460 m (17,910 ft). Peak power output was similar in the hypoxic (10.7 W · kg^{-1}) and normoxic trials (10.9 W · kg^{-1}), whereas average power output was significantly lower in the hypoxic (6.8 W · kg^{-1}) versus the normoxic trial (7.0 W · kg^{-1}). At a simulated altitude of 5,790 m/18,990 ft (McClellan et al. 1990), there were no differences versus normoxia in peak power output (10.6 vs. 10.8 W · kg^{-1}) or mean power output (8.2 vs. 8.3 W · kg^{-1}) during a 30-sec Wingate test (resistance = 7.5% total body weight) performed by 12 men. Similar results were reported in the same study for a 45-sec Wingate test performed at the same resistance; peak power output (10.4 vs. 10.6 W · kg^{-1}) and mean power output (7.0 vs. 7.0 W · kg^{-1}) were not significantly different in hypoxia and normoxia, respectively.

Although the scientific literature is equivocal regarding the effect of altitude on anaerobic performance, mathematical models suggest that running sprint performance is enhanced at altitude. Figure 2.6 on page 38 (Peronnet et al. 1991) shows the estimated running velocity in sprints ranging from 60 m to 400 m at elevations up to 4,000 m (13,120 ft). According to this model, sprint velocity is approximately 1% to 3% faster depending on the sprint event and elevation. Recently, Arsac (2002) proposed a mathematical model that predicts that international-level men are capable of running the 100-m sprint in a time of ≤9.73 sec at 2,000 m (6,560 ft) and ≤9.64 sec at 4,000 m (13,120 ft). This model also predicts that international-level women are capable of running the 100-m sprint in a time of ≤10.70 sec at 2,000 m and ≤10.60 sec at 4,000 m. At present (2003), the world records for the 100-m sprint are 9.78 sec and 10.49 sec for men (Tim Montgomery; Paris, France; 2002) and women (Florence Griffith-Joyner; Seoul, South Korea; 1988), respectively. Predicted improvements in 100-m sprint time have been attributed to a 2% (2,000 m) and 4% (4,000 m) reduction in the cost of overcoming air resistance (Arsac 2002). Interestingly, these mathematical models are supported by the results of the 1968 Mexico City Summer Olympics (2,300 m/7,544 ft), where world records were set in the men's 100 m, 200 m, and 400 m as well as in the 4 × 100-m and 4 × 400-m relays (see the introduction for details, p. xv).

In summary, the effect of acute altitude exposure on anaerobic performance is unclear. Some studies on untrained individuals have shown that peak and average power output are not reduced at altitude. However, other investigations that involved elite athletes using relatively high workloads have demonstrated that average power output may be compromised at altitude, particularly in the later stages of a set of repetitive cycling sprints with minimal recovery. Well-trained and elite athletes should be aware of this and structure their sprint workouts

Table 2.7 Effect of Altitude Exposure on Anaerobic Performance in Untrained Individuals and Trained Athletes

Author	Altitude (m/ft)	Subjects	Performance test	Results
Brosnan et al. 2000	2,100/ 6,890[a]	Female elite cyclists (n = 8) Australian NT	Three sets of 6 × 15-sec sprints (work:recovery ratios were 1:3, 1:2, and 1:1, respectively, for the three sets)	Average power output (W) was similar in hypoxia versus normoxia in the first sprint set but was significantly lower in the second and third sprint set.
Balsom et al. 1994	3,000/ 9,840[b]	Male physical education students (n = 7)	10 × 6-sec cycling sprints (30-sec recovery)	Average power output (W) in sprints 8 and 9 was significantly lower in the hypoxic trial versus the normoxic trial.
Weyand et al. 1999	4,270/ 14,005[a]	Healthy untrained males (n = 4)	15 maximal running sprints lasting from 15 to 180 sec (recovery varied from 10 to 20 min depending on the length of the sprint)	Sprint velocity in hypoxia was similar to that in normoxia in the shorter sprints of 15 sec and 60 sec. Sprint velocity was significantly slower in hypoxia versus normoxia in the sprints of 75 sec and 180 sec.
McClellan et al. 1993	5,460/ 17,910[a]	Healthy untrained males (n = 12)	45-sec Wingate test (resistance = 7.5% TBW)	Peak power output was similar in the hypoxic and normoxic trials. Average power output was significantly lower in the hypoxic versus normoxic trial.
McClellan et al. 1990	5,790/ 18,990[a]	Healthy untrained males (n = 12)	30-sec and 45-sec Wingate test (resistance = 7.5% TBW for both tests)	No differences versus normoxia in peak power output (W · kg^{-1}) or mean power output (W · kg^{-1}) for either the 30-sec or 45-sec Wingate test.

Studies are listed by ascending altitude.
[a] Hypoxic gas mixture.
[b] Hypobaric chamber.
NT = national team; TBW = total body weight; W = watts.

accordingly to ensure optimal anaerobic training while at altitude. Table 2.7 summarizes the findings of investigations that have evaluated the effect of altitude on anaerobic/sprint performance. Chapter 3 ("Performance at Sea Level Following Altitude Training") reviews additional studies describing the effect of long-term altitude training on anaerobic/sprint performance.

Cardiovascular Responses

As described in chapter 2, cardiac output (L · min^{-1}) is the volume of blood ejected by the heart per minute to the peripheral circulation, and is a function of heart rate (bpm) and stroke volume (ml · beat^{-1}). At altitude, changes occur in heart rate, stroke volume, and cardiac output. It is important to note that most of the knowledge we currently have regarding cardiovascular responses at altitude comes from some of the classic altitude studies, which were conducted at relatively high elevations (>3,100 m/10,170 ft) using either untrained or moderately trained subjects. Nevertheless, it is probably safe to assume that similar, but perhaps attenuated, cardiovascular responses occur in well-trained or elite athletes upon exposure to more moderate elevations where altitude training typically takes place.

Heart Rate

Within the first few days of exposure to altitude, heart rate at rest and during submaximal exercise is elevated compared with that at sea level (Grover at al. 1976, 1986; Klausen 1966; Vogel et al. 1967, 1974; Welch 1987), whereas heart rate during maximal exercise at altitude may be similar to or lower than that at sea level (Bouissou et al. 1986; Stenberg et al. 1966). Changes in resting and exercise heart rate that occur in the initial few days at altitude do not appear to be altered after several days of altitude acclimatization (Wolfel et al. 1994). Given that many coaches and athletes use heart rate as a reference of workout intensity, it is imperative that they make the appropriate adjustments to heart rate-based "training zones" in order to attain an optimal training effect and to avoid overworking or overtraining.

Stroke Volume

Stroke volume is the amount of blood ejected by the left ventricle of the heart in a single heartbeat, and is dependent on several factors including total blood volume (plasma volume + red cell volume), left ventricular size, venous return, and cardiac muscle contractility. Within the first few hours of exposure to altitude, stroke volume at rest and during submaximal and maximal exercise is similar to or may be slightly lower than that at sea level (Boutellier and Koller 1981; Hartley et al. 1973; Wagner et al. 1986; Wolfel et al. 1994). However, a more dramatic reduction in stroke volume may occur within two days at altitude at rest (Alexander et al. 1967; Klausen 1966; Moore et al. 1986; Vogel et al. 1967) and during submaximal (Alexander et al. 1967; Grover et al. 1976; Vogel et al. 1967) and maximal exercise (Grover et al. 1976; Horstman et al. 1980; Vogel et al. 1967). This reduction in stroke volume may be due to lower total blood volume secondary to a decrease in plasma volume (Jung et al. 1971; Sawka et al. 2000). Stroke volume may continue to decrease over the course of several days at altitude (Alexander et al. 1967; Vogel et al. 1967; Wolfel et al. 1994).

Cardiac Output

As just discussed, heart rate at rest and during submaximal exercise is increased within the first few days of altitude exposure, whereas stroke volume is markedly reduced at rest and during submaximal and maximal exercise within two days of arriving at altitude. Because of the decrease in stroke volume and despite the concomitant increase in heart rate, cardiac output at rest and during submaximal and maximal exercise is reduced compared with sea level values during the initial two days at altitude (Klausen 1966; Vogel et al. 1967). Cardiac output may continue to decrease over the course of several days at altitude (Alexander et al. 1967; Vogel et al. 1967; Wolfel et al. 1994), an effect that is attributable to the altitude-induced decrement in stroke volume (Alexander et al. 1967; Vogel et al. 1967; Wolfel et al. 1994). Altitude-induced changes in heart rate, stroke volume, and cardiac output are summarized in table 2.8.

Table 2.8 Cardiovascular Responses Upon Acute Exposure to Altitude Versus Sea Level

	Rest	Submaximal exercise	Maximal exercise
Heart rate (bpm)	↑	↑	↔ ↓
Stroke volume (ml · beat^{-1})	↓	↓	↓
Cardiac output (L · min^{-1})	↓	↓	↓

Responses based on an exposure of two to three days at an altitude greater than or equal to 3,100 m/10,170 ft.

↔ = similar; ↑ = increase; ↓ = decrease.

Cardiac Function

Recent preliminary evidence suggests that prolonged endurance exercise at altitude may have adverse acute effects on cardiac function (Shave et al. 2002; Whyte et al. 2002). Eight well-trained male cyclists ($\dot{V}O_2$max = 67.4 ml · kg^{-1} · min^{-1}) performed two laboratory-based 50-mile cycling trials at an intensity equivalent to their normobaric normoxic lactate threshold. One cycling trial was completed in normobaric normoxia, whereas the other was done in normobaric hypoxia (F_IO_2 = 0.15) at a simulated altitude of 2,000 m (6,560 ft). Left ventricular systolic ejection fraction (EF) and fractional shortening (FS) were significantly reduced in the normobaric hypoxic trial, returning to normal 48 hr postexercise. Similar reductions in left ventricular systolic EF and FS were not observed during the normobaric normoxic trial (Whyte et al. 2002). In addition, the hypoxic 50-mile cycling trial resulted in significantly elevated cardiac troponin T (cTnT) indicative of secondary cardiac damage. There was no significant elevation of cTnT in the normoxic 50-mile cycling trial (Shave et al. 2002).

Ventilatory Responses

Upon acute exposure to altitude, one of the initial physiological responses is an increase in pulmonary ventilation ($\dot{V}_E$) in an effort to provide the body's tissues and organs with sufficient oxygen. This increase in $\dot{V}_E$ in response to acute altitude exposure occurs within minutes and is initiated by peripheral chemoreceptors that quickly sense the hypoxic environment and send neural signals to the ventilatory control center in the brain to increase pulmonary ventilation (Dempsey and Forster 1982). Within approximately 30 min, there is an attenuation of the initial $\dot{V}_E$ increase (Easton et al. 1986). However, over the next several hours and days of altitude exposure there is a gradual, time-dependent increase in $\dot{V}_E$ (Asano et al. 1997; Fatemian et al. 2001). This long-term ventilatory acclimatization to altitude is caused by enhanced sensitivity of the peripheral chemoreceptors (Katayama et al. 1999, 2001).

The hypoxic ventilatory response (HVR) is a valid noninvasive method of measuring the sensitivity of the peripheral chemoreceptors to hypoxia. Several studies have shown that HVR increases after intermittent hypoxic exposure (Garcia et al. 2000; Katayama et al. 1998; Levine et al. 1992) or during natural altitude acclimatization (Rivera-Ch et al. 2003; Sato et al. 1992, 1994; White et al. 1987). In terms of athletic performance, an increase in HVR could be beneficial for athletes upon their return to sea level because it could potentially enhance oxygen delivery to the working muscles during the high-intensity phases (that is, "hypoxic" phases) of endurance events. A recent study by Townsend et al. (2002) demonstrated that, compared with prealtitude values, HVR was significantly elevated in trained cyclists and triathletes for two days after they had slept for 20 nights, 8 to 10 hr per night, at a simulated altitude of 2,650 m (8,690 ft). However, it appears that the altitude-induced increase in HVR is lost within a few days after returning to sea level (Sato et al. 1992, 1994; White et al. 1987).

Respiratory and Urinary Water Loss

The maintenance of proper fluid balance is a concern for athletes at sea level, especially for those involved in sports conducted in a hot and humid environment. Proper hydration is even more important for athletes training at altitude. Within the first few days at altitude, there is a tendency toward dehydration due to (1) increased *respiratory* water loss (Kayser 1994) secondary to enhanced pulmonary ventilation (Dempsey and Forster 1982; Laciga and Koller 1976; Moore et al. 1986) and (2) increased *urinary* water loss secondary to downregulation of the renin-angiotensin-aldosterone hormonal mechanism (Hogan et al. 1973; Maher et al. 1975). At moderate altitudes of 2,500 m to 4,300 m (8,200-14,110 ft), respiratory water loss may be as high as 1,900 ml per day in men (Butterfield et al. 1992) and 850 ml per day in women (Mawson et al. 2000). Urinary water loss may average approximately 500 ml per day (Butterfield 1996). Given the potential for performance-limiting dehydration, athletes training at altitude need to maintain fluid balance through regular hydration. They should do this in conjunction with daily workouts as well as during the non-workout period of the day. Fluid intake in the form of water, juices, and carbohydrate-electrolyte drinks should be increased to as much as 4 to 5 L per day to ensure adequate hydration. Caffeinated beverages (coffee, tea, soda), which serve as diuretics, should be avoided or eliminated.

Blood Lactate Response and Acid-Base Balance

The blood lactate response at altitude is often referred to as the "lactate paradox." Reeves et al. (1992) described the lactate paradox as a physiological response in which blood lactate concentration during submaximal and maximal exercise is *increased upon acute altitude exposure* but is *decreased with altitude acclimatization*. The paradox relates to the fact that there is a decrease in submaximal exercise lactate concentration as a result of altitude acclimatization, without an equivalent and concomitant reduction in hypoxic stress. Kayser (1996) hypothesized that the acclimatization-induced reduction in lactate accumulation during submaximal and maximal exercise is mediated by at least two potential mechanisms: (1) a decrease in the maximum substrate flux via aerobic glycolysis due to the reduction in $\dot{V}O_2$max that is experienced at altitude; and (2) alterations in the metabolic control of glycogenolysis and glycolysis at the cellular level, which may be due to changes in sympathetic nervous system response that occur with acclimatization. A recent study (Lundby et al. 2000) showed that among a group of mountaineers, peak blood lactate concentration was lower after one and four weeks of acclimatization at approximately 5,400 m (17,710 ft); but after six weeks of acclimatization this value had returned to a level similar to that observed at sea level and during acute exposure to simulated high altitude. These data suggested that the lactate paradox may be a transient phenomenon that is reversed during a period at altitude lasting six weeks or more.

As described in the lactate paradox model, blood lactate response during submaximal and maximal exercise in the first few hours and days at altitude may be more pronounced than the response during similar exercise at sea level. It is important for coaches and athletes to be aware of this potential physiological limitation as they design individual workouts. High blood lactate levels and accompanying high concentrations of hydrogen ions (H^+) are known to contribute to skeletal muscle fatigue by impairing actin-myosin cross-bridge cycling, reducing the sensitivity of troponin for calcium (Ca^{2+}), and inhibiting the enzyme phosphofructokinase (PFK) and thereby reducing anaerobic energy production via glycolysis (McComas 1996).

In addition, H^+-buffering capacity may be adversely affected in the initial hours and days at altitude. One of the first physiological responses that occurs in humans upon exposure to altitude is an increase in pulmonary ventilation. This higher ventilatory rate results in an increase in CO_2 exhalation and thus a decrease in the partial pressure of carbon dioxide (PCO_2) (Dempsey and Forster 1982; Laciga and Koller 1976). In addition, there is an increase in H^+ removal via hyperventilation that results in an increase in blood pH (Hansen et al. 1967). This acid-base imbalance stimulates the kidneys to increase the rate of renal bicarbonate excretion within hours of arrival at altitude (Hansen et al. 1967). Given that bicarbonate (HCO_3^-) serves as one of the primary metabolic buffers of H^+ produced during exercise, a reduction in bicarbonate may result in a reduced H^+-buffering capacity and thus may adversely affect training, particularly maximal or supramaximal exercise.

Carbohydrate Utilization

Initial studies evaluating the effect of altitude exposure on substrate utilization suggested an increased reliance on lipids for energy (Young et al. 1982, 1987). This finding was supported by the fact that serum free fatty acid and glycerol levels increased upon acute exposure to 4,300 m (14,110 ft) and increased further after 18 days at altitude (Young et al. 1982). This effect was observed both in the resting state and during submaximal exercise. In addition, there was a reduction in muscle glycogen utilization and serum lactate during submaximal exercise following 18 days at altitude (Young et al. 1982). Taken together, these data suggested that acute altitude exposure and subsequent acclimatization promote greater utilization of fat and less dependence on carbohydrate as an energy substrate both at rest and during exercise.

In contrast, more recent studies using very accurate radioactive isotopic tracers have suggested an alternative substrate utilization pattern, that is, that carbohydrate utilization is enhanced and lipid utilization is reduced at altitude (Brooks et al. 1991; Roberts et al. 1996a, 1996b). Within 4 hr of arrival at 4,300 m (14,110 ft), fatty acid consumption of the leg muscles was not different versus sea level values when measured at rest and during submaximal exercise (Roberts et al. 1996a); but glucose rate of appearance and glucose oxidation percent were significantly greater compared with sea level values at rest and during exercise (Roberts et al. 1996b). After 21 days of altitude acclimatization, resting and exercise fatty acid consumption was significantly lower compared with sea level values (Roberts et al. 1996a), whereas the glucose rate of appearance and glucose oxidation percent remained significantly greater compared with sea level values at rest and during exercise (Roberts et al. 1996b). Collectively, these findings suggested that there is an increased dependence on blood glucose, the most oxygen-efficient fuel, as an energy substrate during acute and chronic altitude exposure (Brooks et al. 1991; Roberts et al. 1996b). A couple of recent studies, however, have shown that this substrate utilization pattern may be less dominant in women than in men (Beidleman et al. 2002; Braun et al. 2000).

It is important to note that the scientific investigations on carbohydrate utilization reviewed so far in this section were conducted at an elevation that is impractical—that is, too high for athletes to train at. However, it is possible that the significant increase in glucose utilization at 4,300 m (14,110 ft) reported in several of these studies may occur to a lesser degree at moderate altitude where athletes typically live and train. A recent investigation showed that plasma glucose levels were significantly higher in male and female distance runners ($\dot{V}O_2$max = 64.6 ml $\cdot$ kg^{-1} $\cdot$ min^{-1}) following an interval training session done at 1,800 m (5,905 ft) compared with the same workout at sea level (Niess et al. 2003). The interval training session consisted of 10 × 1,000 m with a 2-min recovery between work intervals. The interval training workout at altitude was done following six days of acclimatization and was run at the same relative intensity as the sea level interval training workout, as determined by the runners' blood lactate levels following each 1,000-m interval. The results of this study provided additional support to the fact that carbohydrate utilization is increased during high-intensity training at moderate altitude (1,800 m/5,905 ft). Therefore, it is recommended that athletes make a concerted effort to replenish their carbohydrate stores through high-glycemic food and drink when training at altitude.

Iron Metabolism

Ferritin, the storage form of iron, is a requisite component of hemoglobin synthesis. Thus it is important for endurance athletes to maintain an adequate level of iron, particularly during periods of increased training. This concern is even more important for athletes training at altitude because of the potential increase in erythropoiesis that occurs in the hypoxic environment. Among elite male swimmers, serum ferritin was significantly decreased by day 13 (60 ng · ml^{-1}) of a three-week training camp at 2,225 m (7,300 ft) compared with prealtitude (80 ng · ml^{-1}), and continued to decline over the remainder of the altitude training camp (46 ng · ml^{-1}) (Roberts and Smith 1992). Similar results were recently reported in a case study of an elite female speedskater who completed a 27-day period during which she lived at 2,700 m (8,855 ft) and trained between 1,400 m (4,590 ft) and 3,000 m (9,840 ft) (Pauls et al. 2002). This suggested that the hypoxic environment of altitude may exacerbate an already high training-induced requirement for iron among well-trained athletes. Stray-Gundersen et al. (1992) reported that altitude-induced erythrocythemia did not take place in endurance athletes who were diagnosed as iron deficient (serum ferritin < 20 ng · ml^{-1} for females, < 30 ng · ml^{-1} for males) prior to completing a four-week altitude training camp at 2,500 m/8,200 ft (figure 2.9). These data suggested that athletes need to normalize and closely monitor their iron status *before* attempting altitude training if they expect to gain an increase in red blood cell mass. In addition, recent studies have shown decrements in endurance performance in iron-depleted, nonanemic athletes, which suggests that low iron per se may negatively affect performance through other iron-dependent physiological processes such as the cytochrome-c oxidase reaction of the electron transport system (Friedmann et al. 2001; Hinton et al. 2000).

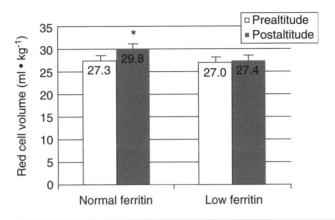

Figure 2.9 Effect of four weeks of living and training at 2,500 m (8,200 ft) on red cell volume in trained female and male distance runners who had normal or low serum ferritin prealtitude. Low serum ferritin was defined as <20 and <30 ng · ml^{-1} for the female and male runners, respectively. *Significantly different versus prealtitude (p < 0.05).

Adapted, by permission, from J. Stray-Gundersen et al., 1992, "Failure of red cell volume to increase to altitude exposure in iron deficient runners," *Medicine and Science in Sports and Exercise* 24(Suppl. 5): S90.

Stress Hormone Response: Cortisol

The glucocorticoid hormone cortisol is released by the adrenal cortex in response to physiological or psychological trauma. One of cortisol's primary physiological functions is the maintenance of blood glucose via gluconeogenesis, for example during prolonged exercise or fasting. Cortisol also works in concert with the catecholamine epinephrine to stimulate lipolysis in the adipose tissue. Abnormally elevated serum cortisol, however, has been associated with skeletal muscle catabolism resulting from extremely high training loads (Aldercruetz et al. 1986; Hakkinen et al. 1989). In addition, elevated serum cortisol has been shown to be related to postexercise immunosuppression, for example neutrophilia, eosinophilia, lymphocythemia and suppression of both natural killer cell and T-cell function (Nieman 1997; Nieman et al. 1994, 1995a, 1995b). Postexercise immunosuppression has been associated with increased rates of upper respiratory tract infection and other minor illnesses that impair an athlete's ability to train and compete (Nieman 1997).

Wilber et al. (2000) recently examined the effect of a five-week training camp at moderate altitude (1,885 m/6,180 ft) on serum cortisol in well-trained endurance athletes. Serum cortisol increased progressively and was significantly elevated at the end of the five-week altitude training period (577 nmol · L^{-1}) relative to 7 to 10 days prealtitude (381 nmol · L^{-1}),

24 to 36 hr after arrival at altitude (411 nmol · L^{-1}), and 18 days after arrival at altitude (422 nmol · L^{-1}). Overall, serum cortisol increased 51% ($p < 0.05$) from prealtitude to the end of the 35-day altitude training camp. Wilber et al. (2000) speculated that the significant increase in serum cortisol observed at the end of the altitude training camp reflected the additive effect of five weeks of hypobaric hypoxia in combination with a progressively increased training load. Serum cortisol increased only 8% (NSD) in the initial 24 to 36 hr at 1,885 m relative to prealtitude values, suggesting that serum cortisol was minimally affected by acute altitude exposure. Throughout the duration of the five-week altitude training camp, however, the serum cortisol response generally reflected the progressive increase in training volume and/or intensity, suggesting that training stress may have affected the cortisol response to a greater degree than hypoxic stress.

Similar findings were reported by Uchakin et al. (1995) in an investigation of the effects of four weeks of living at 2,500 m (8,200 ft) and training at 1,250 m (4,100 ft) in trained runners. The authors reported a postaltitude serum cortisol concentration of 590 nmol · L^{-1}, which was significantly higher than a prealtitude value of 417 nmol · L^{-1}. Likewise, Vasankari et al. (1993) observed that prealtitude serum cortisol (467 nmol · L^{-1}) increased significantly after 14 to 18 days (604 nmol · L^{-1}) at 1,650 m (5,410 ft) in elite male cross-country skiers and biathetes.

In contrast, Rusko et al. (1996) found that serum cortisol was unchanged in well-trained Nordic skiers who trained and raced for 18 to 28 days at an elevation of 1,600 to 1,800 m (5,250-5,905 ft). Similarly, Tsai et al. (1992) reported that serum cortisol was unaltered relative to sea level values in Swedish elite male runners when measured on days 3 and 9 of a 14-day training camp at 2,000 m (6,560 ft). However, cortisol levels for the Swedish runners were significantly higher than those of a group of age- and fitness-matched Kenyan runners who resided and trained at 2,000 m/6,560 ft (Tsai et al. 1992), suggesting a more pronounced hypoxic stress in the native lowlanders (Swedish runners) versus the altitude natives (Kenyan runners). The discrepancy in findings among the various studies of serum cortisol response at moderate altitude may be due to differences in the hypoxic stimulus, that is, differences in the specific altitude, duration of exposure, or both. From a practical standpoint, serum cortisol may serve as a valuable marker for coaches and athletes to use to track training stress, training adaptation, and recovery during altitude training camps.

Immune Function

Many individuals experience upper respiratory tract infections (URTI) or gastrointestinal infections upon exposure to altitude. It has been suggested that this increase in the frequency of URTI and gastrointestinal infections may be due to an altitude-induced suppression of the immune system caused by increased serum levels of stress hormones, such as cortisol (Bailey et al. 1998; Wilber et al. 2000).

At present, there are limited data on the effect of altitude training on the immune response of well-trained athletes. Bailey et al. (1998) evaluated two groups of elite British male distance runners who trained for four weeks at either 1,500 to 2,000 m (4,920-6,560 ft) or 1,640 m (5,380 ft). There was a 50% increase in the frequency of upper respiratory and gastrointestinal tract infections during the altitude training camps in both groups (figure 2.10). In the same study, changes in plasma glutamine (a substrate for lymphocyte and macrophage synthesis) were examined in the runners. Bailey et al. (1998) reported that resting plasma glutamine concentrations (combined data of both groups) decreased 19% after three weeks at altitude ($p < 0.05$), with the greatest decrease observed in two Commonwealth Games medalists who complained of extreme fatigue. It has been suggested that an abnormal decrease in the concentration of plasma glutamine may compromise defense mechanisms against URTI and gastrointestinal infection (Rowbottom et al. 1996). Thus, it appears that altitude training may lead to immunosuppression and an increased risk of URTI and gastrointestinal infection.

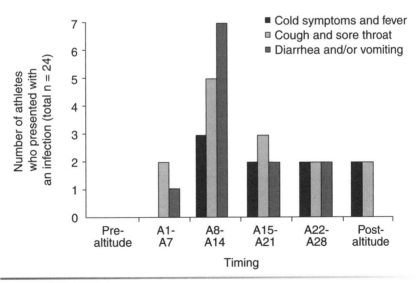

Figure 2.10 The incidence of infectious illness at sea level and during two altitude sojourns (1,500-1,640 m/4,920-5,380 ft). A1-A28 represent days spent at 1,500 to 1,640 m.

Reprinted, by permission, from D. Bailey et al., 1998, "Implications of moderate altitude training for sea-level endurance in elite distance runners," *European Journal of Applied Physiology* 78: 364.

Oxidative Stress

By oxidative stress, we are referring to the production of free radicals, which are molecules or molecular fragments that contain unpaired electrons in their outer orbits (Reif 1992). Oxidative stress has been shown to increase during high-intensity aerobic exercise due to a 10- to 15-fold increase in oxygen consumption in combination with a 1% to 2% electron leakage from the electron transport system and subsequent reduction of molecular oxygen to the superoxide free radical (Alessio 1993; Kanter 1994). Exercise-induced oxidative stress has been implicated in the damage of cellular membranes, increased cellular swelling, decreased cell membrane fluidity, deoxyribonucleic (DNA) damage, and skeletal muscle protein changes, all of which can result in fatigue, delayed onset muscle soreness, increased recovery time, and increased injury rate (Alessio 1993; Pyne 1994).

Oxidative stress may be increased at altitude for several reasons, including low ambient temperatures, increased hypoxia, tissue reperfusion, increased liberation of transition metals (Fe, Mg, Cu) involved in free radical production, a decreased anaerobic threshold, and increased ultraviolet light exposure (Chao et al. 1999; Pfeiffer et al. 1999; Simon-Schnass 1994; Wozniak et al. 2001). A limited number of studies have evaluated the effects of living and training at altitude on oxidative stress in well-trained athletes. Oxidative stress (serum and low-density lipoprotein [LDL]-cholesterol diene conjugation) was not increased in endurance athletes who completed a maximal cycling exercise test in conditions of acute hypoxia (F_iO_2 = ~0.32, P_iO_2 = ~243 mm Hg) compared with normoxia (Peltonen et al. 2001). Similarly, Subudhi et al. (2001) recently reported that oxidative stress markers (malondialdehyde, MDA; lipid hydroperoxides, LOOH) among U.S. national team alpine skiers were not higher versus prealtitude levels after four days of on-snow training at 2,200 m (7,215 ft).

In contrast, well-trained rowers and kayakers demonstrated an increase in oxidative stress (thiobarbituric acid reactive substances, TBARS) after 4, 10, and 18 days of training at 2,000 m (6,560 ft) compared with prealtitude values (Wozniak et al. 2001). Vasankari et al. (1997) reported that oxidative stress (serum diene conjugation) in elite male Nordic skiers was higher at 1,650 m (5,410 ft) versus sea level both before (25%, NSD) and after (30%, $p < 0.05$) a 20-km cross-country ski race. Similar increments in oxidative stress markers have been reported in soldiers who completed field training for 14 days at 2,700 m/8,855 ft (Pfeiffer et al. 1999) and 14 days at approximately 3,000 m/9,840 ft (Chao et al. 1999). Collectively, these studies indicate that long-term living and training at altitude may increase oxidative stress in athletes. It has been suggested that supplementation with the antioxidant vitamin E (400 mg per day for 10 weeks) serves to offset this altitude-induced increase in oxidative stress and may attenuate decrements in aerobic performance at moderate altitude (Simon-Schnass and Pabst 1988).

Sympathetic Nervous System: Epinephrine and Norepinephrine

The sympathetic nervous system is responsible for the regulation of several physiological responses including heart rate, respiratory rate, and substrate utilization. The two primary mediators of the sympathetic nervous system are the catecholamines, epinephrine and norepinephrine. Several studies have demonstrated that the sympathetic nervous system is up-regulated at rest and during exercise in response to acute and chronic exposure to 4,300 m (14,110 ft) as evidenced by increased plasma and urinary levels of epinephrine and norepinephrine (Mazzeo et al. 1991, 1994, 1998, 2001). A recent study has shown similar increases in plasma epinephrine and norepinephrine in male and female distance runners ($\dot{V}O_2$max = 64.6 ml · kg^{-1} · min^{-1}) following a high-intensity interval session (10 × 1,000 m, 2-min recovery) done at moderate altitude (1,800 m/5,905 ft) compared with the same workout at sea level (Niess et al. 2003).

The hypoxia-induced increase in epinephrine is more acute than for norepinephrine, occurring within the first hours of altitude exposure but returning to sea level values by the fifth day at altitude. This acute increase results from the release of epinephrine from the adrenal medulla, which in turn acts on β-adrenergic receptors to preserve blood flow and oxygen delivery to essential tissues and organs via an increased heart rate, stroke volume, and cardiac output (Mazzeo and Reeves 2003; Mazzeo et al. 1991, 1994) (figure 2.11a). The increase in blood lactate concentration during submaximal and maximal exercise that is seen upon acute altitude exposure (i.e., the "lactate paradox") has been associated with the hypoxia-induced increase in epinephrine (Kayser 1996; Lundby et al. 2000). Norepinephrine peaks by the seventh day at altitude and remains elevated throughout the duration of altitude residence (Mazzeo et al. 1991, 1994). This chronic norepinephrine response is caused by increased sympathetic nervous system activity. The resultant increase in plasma norepinephrine acts on α-adrenergic receptors to preserve blood flow and oxygen delivery to essential tissues and organs via increases in ventilatory rate ($\dot{V}_E$), vasoconstriction, vascular pressure, and blood pressure (figure 2.11b)

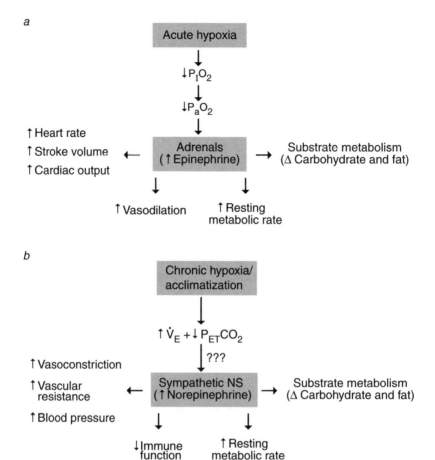

Figure 2.11 Likely mechanism whereby (a) acute hypoxia elicits a strong adrenal medullary response releasing epinephrine into the circulation, and (b) chronic hypoxia elicits an increase in sympathetic nervous system activity resulting in an increase in plasma norepinephrine. P_IO_2 = partial pressure of inspired oxygen; P_aO_2 = partial pressure of oxygen in arterial blood; $\dot{V}_E$ = pulmonary ventilation; $P_{ET}CO_2$ = partial pressure of end-tidal volume CO_2; NS = nervous system; ↑ = increase; ↓ = decrease; Δ = change; ??? = proposed mechanism (inconclusive).

Reprinted, by permission, from R.S. Mazzeo and J.T. Reeves, 2003, "Adrenergic contribution during acclimatization to high altitude: Perspectives from Pikes Peak," *Exercise and Sport Sciences Reviews* 31: 14, 16.

(Mazzeo et al. 1991, 1994; Sevre et al. 2001). This may be a more "energy conservative" approach than the acute hypoxic response, that is, it may be "cheaper" to deliver oxygen to essential tissues and organs via an increase in $\dot{V}_E$ than by increases in heart rate, stroke volume, and cardiac output (Mazzeo and Reeves 2003).

Skeletal Muscle Adenosine Triphosphatases

An adenosine triphosphatase (ATPase) is an enzyme that uses the energy released from the hydrolysis of adenosine triphosphate (ATP) to perform a specific physiological function. Two such ATPases are found within skeletal muscle: sodium potassium-ATPase (Na^+K^+-ATPase) and calcium-ATPase (Ca^{2+}-ATPase). The function of Na^+K^+-ATPase is to restore ionic (Na^+K^+) balance across the sarcolemma and T-tubule within the skeletal muscle cell and thus help regulate membrane excitability and neural transmission. The function of Ca^{2+}-ATPase is to resequester Ca^{2+} from the cytosol back into the sarcoplasmic reticulum of the skeletal muscle cell. Both of these ATPases are very active during skeletal muscle contraction and therefore are important during exercise. Essentially, Na^+K^+-ATPase and Ca^{2+}-ATPase allow skeletal muscle to produce mechanical force quickly and efficiently (Green 2000). At sea level, the concentrations of Na^+K^+-ATPase and Ca^{2+}-ATPase are enhanced via training.

It appears that the enzymatic activity of both Na^+K^+-ATPase and Ca^{2+}-ATPase is reduced upon exposure to high altitude. Studies by Green et al. (2000a, 2000b) have shown significant reductions in Na^+K^+-ATPase and Ca^{2+}-ATPase concentration in mountaineers following a 21-day expedition to Mount Denali (6,194 m/20,315 ft). Whether similar changes occur at less extreme elevations that are more typical of altitudes used by athletes remains to be seen. It is possible that this down-regulation of Na^+K^+-ATPase and Ca^{2+}-ATPase may contribute to the decrements in $\dot{V}O_2$max, endurance performance, and training capacity that have been reported in athletes at moderate altitude.

Body Composition

An individual's body composition may be altered upon exposure to altitude. Initially, there may be a slight reduction in total body weight due to increased respiratory and urinary water loss. Chronic exposure to altitude may lead to a reduction in total body weight resulting from decrements in fat mass and muscle mass (Boyer and Blume 1984; Hoppeler et al. 1990; MacDougall et al. 1991; Rose et al. 1988). These decrements in fat and muscle mass may be a consequence of increases in the basal metabolic rate (Butterfield et al. 1992) and activity level (Kayser 1992) in combination with decreased caloric intake. However, Kayser (1992) has indicated that up to altitudes of approximately 5,000 m (16,400 ft), people can prevent weight loss from fat and muscle by maintaining an adequate and varied caloric intake. This fact is very encouraging to athletes because altitude training camps are typically conducted at elevations well below 5,000 m. Indeed, the postaltitude body composition of Swedish elite male distance runners was not significantly different from their prealtitude value following two weeks of training at 2,000 m/6,560 ft (Svedenhag et al. 1991). Similar findings were reported in Australian elite male track cyclists following three weeks of training at 2,690 m (8,825 ft) (Gore et al. 1998). Nevertheless, athletes training at altitude should make a concerted effort to maintain optimal body weight and body composition. They can accomplish this through proper nutrition and hydration.

Female-Specific Physiological Responses at Altitude

Several studies have evaluated the effects of altitude on women. It should be noted that most of these investigations were conducted at the summit of Pikes Peak, Colorado, at an elevation of 4,300 m (14,110 ft), which is considerably higher than where most athletes conduct altitude training. Nevertheless, the findings of these female-specific studies should be considered relevant for women who live or train at altitude.

When compared with values in age- and fitness-matched men, there was no difference in ventilatory acclimatization (resting ventilation, hypoxic ventilatory response, hypercapnic ventilatory response) among women during a 12-day residence at 4,300 m (Muza et al. 2001). Women had higher levels of the stress hormone cortisol and growth hormone compared with men at simulated sea level during rest and submaximal exercise; however, this pattern was unaffected at a simulated elevation of 2,160 m/7,085 ft (Sandoval and Matt 2002). Likewise, plasma epinephrine and norepinephrine were similar between women and men at rest and during submaximal exercise at simulated sea level and 2,160 m (Sandoval and Matt 2002). As with men, women may experience weight loss at altitude (4,300 m) due to a higher basal metabolic rate, and therefore may need to increase their total caloric intake in order to maintain total body weight and body composition (Mawson et al. 2000).

Carbohydrate utilization is increased at altitude (4,300 m/14,110 ft) compared with sea level, as described earlier in this chapter. However, it appears that this increase in carbohydrate utilization may be attenuated in women. Braun et al. (2000) found that carbohydrate utilization (blood glucose rate of disappearance) in women at rest and during submaximal exercise was not different at sea level and 4,300 m.

Similar results were recently reported by Beidleman et al. (2002) in women upon acute exposure to a simulated elevation of 4,300 m. These findings are in contrast to those from earlier studies on men (Roberts et al. 1996a, 1996b) that showed significantly greater carbohydrate utilization at 4,300 m versus sea level. It has been suggested that the relative abundance of estrogen and progesterone in women may account in part for this gender difference in substrate utilization (Braun et al. 2000).

A number of studies have evaluated the effect of the menstrual cycle on several physiological responses at altitude. Essentially, these investigations have shown that substrate utilization (Beidleman et al. 2002), ventilatory response at rest and during submaximal and maximal exercise (Beidleman et al. 1999; Muza et al. 2001), peak oxygen consumption and endurance exercise performance (Beidleman et al. 1999), plasma epinephrine and norepinephrine response during progressive exercise to exhaustion (Mazzeo et al. 2001), hypoxia-induced erythropoiesis (Reeves et al. 2001), and total energy requirement (Mawson et al. 2000) are not affected by menstrual cycle phase in women residing at 4,300 m.

A recent study by Sandoval and Matt (2003) evaluated the effects of the oral contraceptive pill cycle on physiological responses during moderate exercise at a simulated altitude of 3,500 m (11,480 ft). During the "pill" week of the subjects' monthly oral contraceptive cycle, plasma glucose levels were greater whereas plasma lactate levels were lower at simulated altitude versus sea level during exercise and recovery. These results suggested that elevations in estradiol and progesterone during the pill phase of the oral contraceptive cycle alter lactate and glucose response during exercise at altitude. Based on these findings, the authors concluded that women may perform better in endurance events at altitude during the pill versus non-pill phase of the oral contraceptive cycle (Sandoval and Matt 2003).

To summarize, except for substrate utilization, physiological responses at altitude appear to be similar in women and men, and do not appear to be affected by menstrual cycle phase. Preliminary data suggests that glucose and lactate response during exercise at altitude may be influenced by oral contraceptives.

Sleep Disturbances

Athletes who are exposed to an altitude environment may experience some degree of sleep disturbance. This is particularly true for athletes training at altitude for the first time. Sleep disturbances may be defined as an increase in the number of awakenings throughout the sleep period, periodic labored breathing, and a decrease in slow wave sleep and rapid eye movement (REM) sleep (Goldenberg et al. 1992; Zielinski et al. 2000). Although most of these symptoms have been documented in alpine climbers involved in high-altitude expeditions (e.g., Mount Everest; elevation = 8,852 m/29,035 ft), athletes who are engaged in training at moderate or low altitude may experience milder forms of sleep disturbance. Kinsman et al. (2002) recently reported that sleep apnea was significantly increased during both REM and non-REM sleep in trained cyclists within the first two days of exposure to a simulated altitude of 2,650 m (8,690 ft). Altitude acclimatization helps to reduce the negative effects of sleep disturbance (Goldenberg et al. 1992). Oxygen enrichment of room air has recently been suggested as a way to improve sleep during the initial days of altitude acclimatization (McElroy et al. 2000; West 2002a, 2002b).

Acute Mountain Sickness

Acute mountain sickness (AMS) is an altitude-induced illness with clinical symptoms of headache, nausea, vomiting, and weakness. Although AMS is more prevalent at high altitude (>5,000 m/16,400 ft), it is not uncommon for athletes training at moderate or low altitude to experience mild AMS. In addition, it has been shown that aerobic exercise performed during the initial 6 to 10 hr at simulated high altitude (4,800 m/15,745 ft) exacerbates the severity and incidence of AMS (Roach et al. 2000). The prescription drug acetazolamide has been shown to be effective in reducing the symptoms of AMS, primarily by increasing arterial oxygen levels and reducing peripheral edema (Bradwell et al. 1992). However, a recent study reported that use of acetazolamide prior to exercise at approximately 3,250 m (10,660 ft) resulted in significantly lower $\dot{V}O_2$max, maximal power output, and exercise duration compared to a placebo drug (Garske et al. 2003). Nonprescription aspirin has also been shown to be effective in reducing some of the symptoms of AMS. Recent research has shown that prealtitude treatment with ginkgo biloba is effective in both reducing the severity of AMS (Gertsch et al. 2002) and preventing AMS (Roncin et al. 1996), possibly because of its antioxidant and anti-inflammatory properties (Gertsch et al. 2002). Physical rest is considered the principal nonpharmacological treatment for moderate AMS (Bartsch 1992).

Sickle Cell Trait and Exercise at Altitude

Sickle cell *trait* is an inherited blood condition in which some red blood cells tend to take on a sickle shape, thereby impairing the red blood cell's ability to bind oxygen molecules. Sickle cell trait does not usually result in anemia. In contrast, sickle cell *anemia* is a chronic inherited anemia in which a large proportion of red blood cells tend to have a sickle shape. Sickle cell trait is generally benign and is found in 8% to 10% of African Americans; it is also found in American whites, but rarely (Roth 1985). Most people with sickle cell trait lead normal and productive lives.

However, individuals with sickle cell trait who exercise vigorously at altitude may be predisposed to a minor risk of splenic infarct (Eichner 1986). A splenic infarct is characterized by the destruction of cells in the spleen due to blockage of blood flow. Cases of splenic infarct have been reported in young men with sickle cell trait who lifted weights, jogged, or played basketball at elevations under 3,050 m/10,000 ft (Lane and Githens 1985; Nichols 1968). The primary clinical symptom in these cases was severe left upper-quadrant abdominal pain that was experienced within 48 hr after arrival at altitude. All of these men recovered shortly thereafter and experienced no additional symptoms or complications. The risk of splenic infarct appears to be restricted to a relatively small subpopulation of individuals with sickle cell trait (Eichner 1986). Nevertheless, athletes with sickle cell trait who compete or train at altitude should be aware of this risk and take appropriate precautions. The recommendation is that athletes with sickle cell trait allow additional days for acclimatization and restrict their initial workouts to low- to moderate-intensity training. In addition, they should make a concerted effort to stay well hydrated.

Cognitive Function

It is well documented that cognitive function is impaired at extremely high elevations (>7,000 m/22,960 ft). However, it appears that cognitive function is not affected at simulated elevations below 3,050 m/10,000 ft (Green and Morgan 1985; Kelman et al. 1969). Although exposure to moderate altitude does not appear to affect the performance of learned tasks (Pearson and Neal 1970), it may increase the amount of time required to learn a specific task (Ernsting 1978). For a comprehensive review on the effect of altitude on mood, behavior, and cognitive function, the reader is referred to Bahrke and Shukitt-Hale (1993).

Summary

As a consequence of reductions in P_IO_2 and P_aO_2, arterial oxyhemoglobin saturation is reduced at altitude during submaximal and maximal exercise. Reductions in S_aO_2 are more pronounced in well-trained athletes than in untrained individuals. Maximal oxygen consumption is significantly reduced by 2% to 29% upon exposure to elevations ranging from 580 m to 4,300 m (1,900-14,110 ft). There is great individual variability in the decrement in $\dot{V}O_2$max. Decrements in maximal oxygen consumption are due primarily to the hypoxic-induced reduction in S_aO_2. Aerobic performance and training capacity are negatively affected upon exposure to elevations ranging from 580 m to 5,200 m (1,900-17,055 ft). Performance decrements range from 2% to 34% depending on the altitude and are associated with altitude-induced reductions in S_aO_2 and $\dot{V}O_2$max. These physiological limitations may force athletes to reduce their daily training volume, training intensity, or both and to modify their competition strategy from what they normally do at sea level. Altitude-induced changes in heart rate, hydration status, acid-base balance, carbohydrate utilization, iron metabolism, and immune function may also require athletes to modify their training regimen at altitude. With the exception of substrate utilization, physiological responses at altitude appear to be similar in women versus men, and do not appear to be affected by menstrual cycle phase. It is important for coaches and athletes to be knowledgeable about these physiological responses and limitations in order to achieve an optimal training effect and avoid overreaching or overtraining. Practical recommendations for dealing with these physiological responses and potential limitations at altitude are summarized in chapter 7, "Recommendations and Guidelines."

References

Adams, W.C., E.M. Bernauer, D.B. Dill, and J.B. Bomar. 1975. Effects of equivalent sea-level and altitude training on $\dot{V}O_2$max and running performance. *Journal of Applied Physiology* 39: 262-266.

Aldercreutz, H., M. Harkonen, K. Kuoppasalmi, H. Naveri, H. Huhtamieni, H. Tikkanen, K. Remes, A. Dessypris, and J. Karvonen. 1986. Effect of training on plasma anabolic and catabolic steroid hormones and their responses during physical exercise. *International Journal of Sports Medicine* 7 (Suppl. 1): 27-28.

Alessio, H.M. 1993. Exercise-induced oxidative stress. *Medicine and Science in Sports and Exercise* 25: 218-224.

Alexander, J.K., L.H. Hartley, M. Modelski, and R.F. Grover. 1967. Reduction of stroke volume in man following ascent to 3100 m altitude. *Journal of Applied Physiology* 23: 849-858.

Anselme, F., C. Caillaud, I. Couret, M. Rossi, and C. Prefaut. 1994. Histamine and exercise-induced hypoxemia in highly trained athletes. *Journal of Applied Physiology* 76: 127-132.

Arsac, L.M. 2002. Effects of altitude on the energetics of human best performances in 100 m running: A theoretical analysis. *European Journal of Applied Physiology* 87: 78-84.

Asano, K., R.S. Mazzeo, R.E. McCullough, E.E. Wolfel, and J.T. Reeves. 1997. Relation of sympathetic activation to ventilation in man at 4,300 m altitude. *Aviation, Space, and Environmental Medicine* 68: 104-110.

Bahrke, M.S., and B. Shukitt-Hale. 1993. Effects of altitude on mood, behaviour and cognitive functioning. *Sports Medicine* 16: 97-125.

Bailey, D.M., B. Davies, L. Romer, L. Castell, E. Newsholme, and G. Gandy. 1998. Implications of moderate altitude training for sea-level endurance in elite distance runners. *European Journal of Applied Physiology* 78: 360-368.

Balsom, P.D., G.C. Gaitanos, B. Ekblom, and B. Sjodin. 1994. Reduced oxygen availability during high intensity intermittent exercise impairs performance. *Acta Physiologica Scandinavica* 152: 279-285.

Banchero, N., F. Sime, D. Penaloza, J. Cruz, R. Gamboa, and E. Marticorena. 1966. Pulmonary pressure, cardiac output, and arterial oxygen saturation during exercise at high altitude and at sea level. *Circulation* 33: 249-262.

Bartsch, P. 1992. Treatment of high altitude diseases without drugs. *International Journal of Sports Medicine* 13 (Suppl. 1): S71-S74.

Bassett, D.R., C.R. Kyle, L. Passfield, J.P. Broker, and E.R. Burke. 1999. Comparing cycling world records, 1967-1996: Modeling with empirical data. *Medicine and Science in Sports and Exercise* 31: 1665-1676.

Beidleman, B.A., P.R. Rock, S.R. Muza, C.S. Fulco, V.A. Forte, and A. Cymerman. 1999. Exercise $\dot{V}_E$ and physical performance at altitude are not affected by menstrual cycle phase. *Journal of Applied Physiology* 86: 1519-1526.

Beidleman, B.A., P.R. Rock, S.R. Muza, C.S. Fulco, L.L. Gibson, G.H. Kamimori, and A. Cymerman. 2002. Substrate oxidation is altered in women during exercise upon acute altitude exposure. *Medicine and Science in Sports and Exercise* 34: 430-437.

Billat, V.L., P.M. Lepretre, R.P. Heubert, J.P. Koralsztein, and F.P. Gazeau. 2003. Influence of acute hypoxia on time to exhaustion at v$\dot{V}O_2$max in unacclimatized runners. *International Journal of Sports Medicine* 24: 9-14.

Bouissou, P., F. Peronnet, G. Brisson, R. Helie, and M. Ledoux. 1986. Metabolic and endocrine responses to graded exercise under acute hypoxia. *European Journal of Applied Physiology* 55: 290-294.

Boutellier, U., and E.A. Koller. 1981. Propranolol and the respiratory, circulatory and ECG responses to high altitude. *European Journal of Applied Physiology* 46: 105-119.

Boyer, S.J., and F.D. Blume. 1984. Weight loss and changes in body composition at high altitude. *Journal of Applied Physiology* 57: 1580-1585.

Bradwell, A.R., A.D. Wright, M. Winterborn, and C. Imray. 1992. Acetazolamide and high altitude diseases. *International Journal of Sports Medicine* 13 (Suppl. 1): S63-S64.

Braun, B., J.T. Mawson, S.R. Muza, S.B. Dominick, G.A. Brooks, M.A. Horning, P.B. Rock, L.G. Moore, R.S. Mazzeo, S.C. Ezeji-Okoye, and G.E. Butterfield. 2000. Women at altitude: Carbohydrate utilization during exercise at 4,300 m. *Journal of Applied Physiology* 88: 246-256.

Brooks, G.A., G.E. Butterfield, R.R. Wolfe, B.M. Groves, R.S. Mazzeo, J.R. Sutton, E.E. Wolfel, and J.T. Reeves. 1991. Increased dependence on blood glucose after acclimatization to 4,300 m. *Journal of Applied Physiology* 70: 919-927.

Brosnan, M.J., D.T. Martin, A.G. Hahn, C.T. Gore, and J.A. Hawley. 2000. Impaired interval exercise responses in elite female cyclists at moderate simulated altitude. *Journal of Applied Physiology* 89: 1819-1824.

Buskirk, E.R., J. Kollias, R.F. Akers, E.K. Prokop, and E.P. Reategui. 1967a. Maximal performance at altitude and on return from altitude in conditioned runners. *Journal of Applied Physiology* 23: 259-266.

Buskirk, E.R., J. Kollias, E. Picon-Reatigue, R. Akers, E. Prokop, and P. Baker. 1967b. Physiology and performance of track athletes at various altitudes in the United States and Peru. In *The International Symposium on the Effects of Altitude on Physical Performance*. Chicago: The Athletic Institute.

Butterfield, G.E. 1996. Maintenance of body weight at altitude. In *Nutritional needs in cold and high-altitude environments*, edited by B.M. Marriott and S.J. Carlson, pp. 357-378. Washington, DC: Committee on Military Nutrition Research.

Butterfield, G.E., J. Gates, S. Flemming, G.A. Brooks, J.R. Sutton, and J.T. Reeves. 1992. Increased energy intake minimizes weight loss in men at high altitude. *Journal of Applied Physiology* 72: 1741-1748.

Capelli, C., and P.E. di Prampero. 1995. Effects of altitude on top speeds during 1 h unaccompanied cycling. *European Journal of Applied Physiology* 71: 469-471.

Chao, W.-H., E.W. Askew, D.E. Roberts, S.M. Wood, and J.B. Perkins. 1999. Oxidative stress in humans during work at moderate altitude. *Journal of Nutrition* 129: 2009-2012.

Chapman, R.F., M. Emery, and J.M. Stager. 1999. Degree of arterial desaturation in normoxia influences $\dot{V}O_2$max decline in mild hypoxia. *Medicine and Science in Sports and Exercise* 31: 658-663.

Chapman, R.F., J. Stray-Gundersen, and B.D. Levine. 1998. Individual variation in response to altitude training. *Journal of Applied Physiology* 85: 1448-1456.

Chick, T.W., D.M. Stark, and G.H. Murata. 1993. Hyperoxic training increases work capacity after maximal training at moderate altitude. *Chest* 104: 1759-1762.

Daniels, J., and N. Oldridge. 1970. The effects of alternate exposure to altitude and sea level in world-class middle-distance runners. *Medicine and Science in Sports* 2: 107-112.

Dempsey, J.A. 1986. Is the lung built for exercise? *Medicine and Science in Sports and Exercise* 18: 143-155.

Dempsey, J.A., and H.V. Forster. 1982. Mediation of ventilatory adaptations. *Physiological Reviews* 62: 262-346.

Dempsey, J.A., P.E. Hanson, and K.S. Henderson. 1984. Exercise induced arterial hypoxemia in healthy persons at sea level. *Journal of Physiology* 355: 161-175.

Dempsey, J.A., and P.D. Wagner. 1999. Exercise-induced arterial hypoxemia. *Journal of Applied Physiology* 87:1997-2006.

Derchak, P.A., J.M. Stager, D.A. Tanner, and R.F. Chapman. 2000. Expiratory flow limitation confounds ventilatory response during exercise in athletes. *Medicine and Science in Sports and Exercise* 32: 1873-1879.

Dill, D.B., and W.C. Adams. 1971. Maximal oxygen uptake at sea level and at 3,090-m altitude in high school champion runners. *Journal of Applied Physiology* 30: 854-859.

Easton, P.A., L.J. Slykerman, and N.R. Anthonisen. 1986. Ventilatory response to sustained hypoxia in normal adults. *Journal of Applied Physiology* 61: 906-911.

Eichner, E.R. 1986. Sickle cell trait, exercise, and altitude. *Physician and Sportsmedicine* 14: 144-157.

Ernsting, J. 1978. The 10th Annual Harry G. Armstrong Lecture: Prevention of hypoxia—acceptable compromises. *Aviation Space and Environmental Medicine* 49: 495-502.

Fatemian, M., D.Y. Kim, M.J. Poulin, and P.A. Robbins. 2001. Very mild exposure to hypoxia for 8 hr can induce ventilatory acclimatization in humans. *Pflugers Archiv* 441: 840-843.

Faulkner, J.A., J.T. Daniels, and B. Balke. 1967. Effects of training at moderate altitude on physical performance capacity. *Journal of Applied Physiology* 23: 85-89.

Faulkner, J.A., J. Kollias, C.B. Favour, E.R. Buskirk, and B. Balke. 1968. Maximum aerobic capacity and running performance at altitude. *Journal of Applied Physiology* 24: 685-691.

Friedmann, B., E. Weller, H. Mairbaurl, and P. Bartsch. 2001. Effects of iron repletion on blood volume and performance capacity in young athletes. *Medicine and Science in Sports and Exercise* 33: 741-746.

Fulco, C.S., P.D. Rock, and A. Cymerman. 1998. Maximal and submaximal exercise performance at altitude. *Aviation Space and Environmental Medicine* 69: 793-801.

Gale, G.E., J.R. Torre-Bueno, R.E. Moon, H.A. Saltzman, and P.D. Wagner. 1985. Ventilation-perfusion inequality in normal humans during exercise at sea level and simulated altitude. *Journal of Applied Physiology* 58: 978-988.

Garcia, N., S.R. Hopkins, and F.L. Powell. 2000. Effects of intermittent hypoxia on the isocapnic hypoxic ventilatory response and erythropoiesis in humans. *Respiratory Physiology* 123: 39-49.

Garske, L.A., M.G. Brown, and S.C. Morrison. 2003. Acetazolamide reduces exercise capacity and increases leg fatigue under hypoxic conditions. *Journal of Applied Physiology* 94: 991-996.

Gertsch, J.H., T.B. Seto, J. Mor, and J. Onopa. 2002. *Ginkgo biloba* for the prevention of severe acute mountain sickness (AMS) starting one day before rapid ascent. *High Altitude Medicine and Biology* 3: 29-37.

Goldenberg, F., J.P. Richalet, I. Onnen, and A.M. Antezana. 1992. Sleep apneas and high altitude newcomers. *International Journal of Sports Medicine* 13 (Suppl. 1): S34-S36.

Gore, C.J., N.P. Craig, A.G. Hahn, A.J. Rice, P.C. Bourdon, S.R. Lawrence, C.B.V. Walsh, P.G. Barnes, R. Parisotto, D.T. Martin, and D.B. Pyne. 1998. Altitude training at 2690m does not increase total haemoglobin mass or sea level $\dot{V}O_2$max in world champion track cyclists. *Journal of Science and Medicine in Sport* 1: 156-170.

Gore, C.J., A.G. Hahn, G.C. Scroop, D.B. Watson, K.I. Norton, R.J. Wood, D.P. Campbell, and D.L. Emonson. 1996. Increased arterial desaturation in trained cyclists during maximal exercise at 580 m altitude. *Journal of Applied Physiology* 80: 2204-2210.

Gore, C.J., S.C. Little, A.G. Hahn, G.C. Scroop, K.I. Norton, P.C. Bourdon, S.M. Woolford, J.D. Buckley, T. Stanef, D.P. Campbell, D.B. Watson, and D.L. Emonson. 1997. Reduced performance of male and female athletes at 580 m altitude. *European Journal of Applied Physiology* 75: 136-143.

Green, H.J. 2000. Altitude acclimatization, training and performance. *Journal of Science and Medicine in Sport* 3: 299-312.

Green, H.J., B. Roy, S. Grant, M. Burnett, R. Tupling, C. Otto, A. Pipe, and D. McKenzie. 2000a. Downregulation in muscle Na$^+$K$^+$-ATPase following a 21-day expedition to 6,194 m. *Journal of Applied Physiology* 88: 634-640.

Green, H.J., B. Roy, S. Grant, M.R. Tupling, C. Otto, A. Pipe, D. McKenzie, and J. Ouyang. 2000b. Effects of a 21-day expedition to 6194 m on human skeletal muscle SR Ca^{2+}-ATPase. *High Altitude Medicine and Biology* 1: 301-310.

Green, R.G., and D.R. Morgan. 1985. The effects of mild hypoxia on a logical reasoning task. *Aviation Space and Environmental Medicine* 56: 1004-1008.

Grover, R.F. 1978. Adaptation to high altitude. In *Environmental stress: Individual human adaptations*, edited by L.J. Folinsbee, J.A. Wagner, J.F. Borgia, B.L. Drinkwater, J.A. Gliner, and J.F. Bedi. New York: Academic Press.

Grover, R.F. 1979. Physiological adaptations to high altitude. In *Sports medicine and physiology*, edited by R.H. Strauss, p. 333. Philadelphia: W.B. Saunders.

Grover, R.F., J.T. Reeves, J.T. Maher, R.E. McCullough, J.C. Cruz, J.C. Denniston, and A. Cymerman. 1976. Maintained stroke volume but impaired arterial oxygenation in man at high altitude with supplemental CO$_2$. *Circulation Research* 38: 391-396.

Grover, R.F., J.V. Weil, and J.T. Reeves. 1986. Cardiovascular adaptation to exercise at high altitude. In *Exercise and sport sciences reviews*, edited by K.B. Pandolf, pp. 269-302. New York: Macmillan.

Hahn, A.G., and C.J. Gore. 2001. The effect of altitude on cycling performance. *Sports Medicine* 31: 533-557.

Hakkinen, K., K.L. Keskinen, M. Alen, P.V. Komi, and H. Kauhanen. 1989. Serum hormone concentrations during prolonged training in elite endurance-trained and strength-trained athletes. *European Journal of Applied Physiology* 59: 233-238.

Hannon, J.P. 1978. Comparative altitude adaptability of young men and women. In *Environmental stress: Individual human adaptations*, edited by L.J. Folinsbee, J.A. Wagner, J.F. Borgia, B.L. Drinkwater, J.A. Gliner, and J.F. Bedi, pp. 335-350. New York: Academic Press.

Hansen, J.R., G.P. Stetler, and J.A. Vogel. 1967. Arterial pyruvate, lactate, pH and PCO$_2$ during work at sea level and high altitude. *Journal of Applied Physiology* 24: 523-530.

Hartley, L.H., J.A. Vogel, and M. Landowne. 1973. Central, femoral, and brachial circulation during exercise in hypoxia. *Journal of Applied Physiology* 34: 87-90.

Haymes, E.M., and C.L. Wells. 1986. *Environment and human performance*. Champaign, IL: Human Kinetics.

Hinton, P.S., C. Giordano, T. Brownlie, and J.D. Haas. 2000. Iron supplementation improves endurance after training in iron-depleted, nonanemic women. *Journal of Applied Physiology* 88: 1103-1111.

Hogan, R.P., T.A. Kotchen, A.E. Boyd III, and L.H. Hartley. 1973. Effect of altitude on renin-aldosterone system and metabolism of water and electrolytes. *Journal of Applied Physiology* 35: 385-390.

Hoppeler, H., E. Kleinert, C. Schlegel, E. Claassen, H. Howald, S.R. Kayar, and P. Cerretelli. 1990. Morphologic adaptations of human skeletal muscle to chronic hypoxia. *International Journal of Sports Medicine* 11 (Suppl. 1): S3-S9.

Horstman, D., R. Weiskopf, and R.E. Jackson. 1980. Work capacity during 3-week sojourn at 4300 m: Effects of relative polycythemia. *Journal of Applied Physiology* 49: 311-318.

Huang, S.Y., J.K. Alexander, R.F. Grover, J.T. Maher, R.E. McCullough, R.G. McCullough, L.G. Moore, J.B. Sampson, J.V. Weil, and J.T. Reeves. 1984. Hypocapnia and sustained hypoxia blunt ventilation on arrival at high altitude. *Journal of Applied Physiology* 56: 602-606.

Jensen, K., T.S. Nielsen, A. Fiskestrand, J.O. Lund, N.J. Christensen, and N.H. Sechar. 1993. High-altitude training does not increase maximal oxygen uptake or work capacity at sea level in rowers. *Scandinavian Journal of Medicine and Science in Sports* 3: 256-262.

Jung, R.C., D.B. Dill, R. Horton, and S.M. Horvath. 1971. Effects of age on plasma aldosterone levels and hemoconcentration at altitude. *Journal of Applied Physiology* 31: 593-597.

Kanter, M.M. 1994. Free radicals, exercise, and antioxidant supplementation. *International Journal of Sport Nutrition* 4: 205-220.

Katayama, K., Y. Sato, K. Ishida, S. Mori, and M. Miyamura. 1998. The effects of intermittent exposure to hypoxia during endurance exercise training on the ventilatory responses to hypoxia and hypercapnia in humans. *European Journal of Applied Physiology* 78: 189-194.

Katayama, K., Y. Sato, Y. Morotome, N. Shima, K. Ishida, S. Mori, and M. Miyamura. 1999. Ventilatory chemosensitive adaptations to intermittent hypoxic exposure with endurance training and detraining. *Journal of Applied Physiology* 86: 1805-1811.

Katayama, K., Y. Sato, Y. Morotome, N. Shima, K. Ishida, S. Mori, and M. Miyamura. 2001. Intermittent hypoxia increases ventilation and S_aO_2 during hypoxic exercise and hypoxic chemosensitivity. *Journal of Applied Physiology* 90: 1431-1440.

Kayser, B. 1992. Nutrition and high altitude exposure. *International Journal of Sports Medicine* 13 (Suppl. 1): S129-S132.

Kayser, B. 1994. Nutrition and energetics of exercise at altitude. *Sports Medicine* 17: 309-323.

Kayser, B. 1996. Lactate during exercise at high altitude. *European Journal of Applied Physiology* 74: 195-205.

Kelman, G.R., T.J. Crow, and A.E. Bursill. 1969. Effect of mild hypoxia on mental performance assessed by a test of selective attention. *Aerospace Medicine* 40: 301-303.

Kinsman, T.A., A.G. Hahn, C.J. Gore, B.R. Wilsmore, D.T. Martin, and C.-M. Chow. 2002. Respiratory events and periodic breathing in cyclists sleeping at 2,650-m simulated altitude. *Journal of Applied Physiology* 92: 2114-2118.

Klausen, K. 1966. Cardiac output in man in rest and work during and after acclimatization to 3800 m. *Journal of Applied Physiology* 21: 609-616.

Koistinen, P., T. Takala, V. Martikkala, and J. Leppaluoto. 1995. Aerobic fitness influences the response of maximal oxygen uptake and lactate threshold in acute hypobaric hypoxia. *International Journal of Sports Medicine* 26: 78-81.

Kyle, C.R., and D.R. Bassett. 2003. The cycling world hour record. In *High-tech cycling* (2nd ed.), edited by E.R. Burke, pp. 175-196. Champaign, IL: Human Kinetics.

Laciga, P., and E.A. Koller. 1976. Respiratory, circulatory, and ECG changes during acute exposure to altitude. *Journal of Applied Physiology* 41: 159-167.

Lane, P.A., and J.H. Githens. 1985. Splenic syndrome at mountain altitudes in sickle cell trait. Its occurrence in nonblack persons. *Journal of the American Medical Association* 253: 2251-2254.

Lawler, J., S.K. Powers, and D. Thompson. 1988. Linear relationship between $\dot{V}O_2$max and $\dot{V}O_2$max decrement during exposure to acute hypoxia. *Journal of Applied Physiology* 64: 1486-1492.

Levine, B.D., D.B. Friedman, K. Engfred, B. Hanel, M. Kjaer, P.S. Clifford, and N.H. Secher. 1992. The effect of normoxic or hypobaric hypoxic endurance training on the hypoxic ventilatory response. *Medicine and Science in Sports and Exercise* 24: 769-775.

Levine, B.D., and J. Stray-Gundersen. 1997. "Living high-training low": Effect of moderate-altitude acclimatization with low-altitude training on performance. *Journal of Applied Physiology* 83: 102-112.

Lundby, C., B. Saltin, and G. van Hall. 2000. The 'lactate paradox,' evidence for a transient change in the course of acclimatization to severe hypoxia in lowlanders. *Acta Physiologica Scandinavica* 170: 265-269.

MacDougall, J.D., H.J. Green, J.R. Sutton, G. Coates, A. Cymerman, P. Young, and C.S. Houston. 1991. Operation Everest II: Structural adaptations in skeletal muscle in response to extreme simulated altitude. *Acta Physiologica Scandinavica* 142: 421-427.

Maher, J.T., L.G. Jones, L.H. Hartley, G.H. Williams, and L.I. Rose. 1975. Aldosterone dynamics during graded exercise at sea level and high altitude. *Journal of Applied Physiology* 39: 18-22.

Marieb, E.N. 1992. *Human anatomy and physiology* (2nd ed.). Redwood City, CA: Benjamin Cummings.

Mawson, J.T., B. Braun, P.B. Rock, L.G. Moore, R. Mazzeo, and G.E. Butterfield. 2000. Women at altitude: Energy requirements at 4,300 m. *Journal of Applied Physiology* 88: 272-281.

Mazzeo, R.S., P.R. Bender, G.A. Brooks, G.E. Butterfield, B.M. Groves, J.R. Sutton, E.E. Wolfel, and J.T. Reeves. 1991. Arterial catecholamine responses during exercise with acute and chronic high altitude exposure. *American Journal of Physiology: Endocrinology and Metabolism* 261: E419-E424.

Mazzeo, R.S., J.D. Carroll, G.E. Butterfield, B. Braun, P.B. Rock, E.E. Wolfel, S. Zamudio, and L.G. Moore. 2001. Catecholamine response to α-adrenergic blockade during exercise in women acutely exposed to altitude. *Journal of Applied Physiology* 90: 121-126.

Mazzeo, R.S., A. Child, G.E. Butterfield, J.T. Mawson, S. Zamudio, and L.G. Moore. 1998. Catecholamine response during 12 days of high altitude exposure (4,300 m) in women. *Journal of Applied Physiology* 84: 1151-1157.

Mazzeo, R.S., and J.T. Reeves. 2003. Adrenergic contribution during acclimatization to high altitude. *Exercise and Sport Sciences Reviews* 31: 13-18.

Mazzeo, R.S., E.E. Wolfel, G.E. Butterfield, and J.T. Reeves. 1994. Sympathetic responses during 21 days at high altitude (4,300 m) as determined by urinary and arterial catecholamines. *Metabolism* 43: 1226-1232.

McClellan, T.M., S.S. Cheung, and M.R. Meunier. 1993. The effect of normocapnic hypoxia and the duration of exposure to hypoxia on supramaximal exercise performance. *European Journal of Applied Physiology* 66: 409-414.

McClellan, T.M., M.F. Kavanaugh, and I. Jacobs. 1990. The effect of hypoxia on performance during 30 s or 45 s of supramaximal exercise. *European Journal of Applied Physiology* 60: 155-161.

McComas, A.J. 1996. *Skeletal muscle form and function*. Champaign, IL: Human Kinetics.

McElroy, M.K., A. Gerard, F.L. Powell, G.K. Prisk, N. Sentse, S. Holverda, and J.B. West. 2000. Nocturnal O_2 enrichment of room air at high altitude increases daytime O_2 saturation without changing control of ventilation. *High Altitude Medicine and Biology* 1: 197-206.

Moore, L.G., A. Cymerman, S.Y. Huang, R.E. McCullough, R.G. McCullough, P.B. Rock, A.J. Young, P.M. Young, D. Bloedow, J.V. Weil, and J.T. Reeves. 1986. Propranolol does not impair exercise oxygen uptake in normal man at high altitude. *Journal of Applied Physiology* 61: 1935-1941.

Morris, D.M., J.T. Kearney, and E.R. Burke. 2000. The effects of breathing supplemental oxygen during altitude training on cycling performance. *Journal of Science and Medicine in Sport* 3: 165-175.

Muza, S.R., P.B. Rock, C.S. Fulco, S. Zamudio, B. Braun, A. Cymerman, G.E. Butterfield, and L.G. Moore. 2001. Women at altitude: Ventilatory acclimatization at 4,300 m. *Journal of Applied Physiology* 91: 1791-1799.

Nichols, S.D. 1968. Splenic and pulmonary infarction in a Negro athlete. *Rocky Mountain Medical Journal* 65: 49-50.

Nieman, D. 1997. Immune response to heavy exertion. *Journal of Applied Physiology* 82: 1385-1394.

Nieman, D.C., J.C. Ahle, D.A. Henson, B.J. Warren, J. Suttles, J.M. Davis, K.S. Buckley, S. Simandle, D.E. Butterworth, O.R. Fagoaga, and S.L. Nehlsen-Cannarella. 1995a. Indomethacin does not alter the natural killer cell response to 2.5 hours of running. *Journal of Applied Physiology* 79: 748-755.

Nieman, D.C., A.R. Miller, D.A. Henson, B.J. Warren, G. Gusewitch, R.L. Johnson, J.M. Davis, D.E. Butterworth, J.L. Herring, and S.L. Nehlsen-Cannarella. 1994. Effects of high- versus moderate-intensity exercise on circulating lymphocyte subpopulations and proliferative response. *International Journal of Sports Medicine* 15: 199-206.

Nieman, D.C., S. Simandle, D.A. Henson, B.J. Warren, J. Suttles, J.M. Davis, K.S. Buckley, J.C. Ahle, D.E. Butterworth, O.R. Fagoaga, and S.L. Nehlsen-Cannarella. 1995b. Lymphocyte proliferation response to 2.5 hours of running. *International Journal of Sports Medicine* 16: 406-410.

Niess, A.M., E. Fehrenbach, G. Strobel, K. Roecker, E.M. Schneider, J. Buergler, S. Fuss, R. Lehmann, H. Northoff, and H.-H. Dickhuth. 2003. Evaluation of stress response to interval training at low and moderate altitudes. *Medicine and Science in Sports and Exercise* 35: 263-269.

Olds, T. 1992. The optimal altitude for cycling performance: A mathematical model. *Excel* 8: 155-159.

Olds, T. 2001. Modelling human locomotion: Applications to cycling. *Sports Medicine* 31: 497-509.

Pauls, D.W., H. van Duijnhoven, and J. Stray-Gundersen. 2002. Iron insufficient erythropoiesis at altitude—speed skating. *Medicine and Science in Sports and Exercise* 34 (Suppl. 5): S252.

Pearson, R.G., and G.L. Neal. 1970. Operator performance as a function of drug, hypoxia, individual and task factors. *Aerospace Medicine* 41: 154-158.

Peltonen, J.E., H.O. Tikkanen, J.J. Ritola, M. Ahotupa, and H.K. Rusko. 2001. Oxygen uptake response during maximal cycling in hyperoxia, normoxia and hypoxia. *Aviation Space and Environmental Medicine* 72: 904-911.

Peronnet, F., G. Thibault, and D.-L. Cousineau. 1991. A theoretical analysis of the effect of altitude on running performance. *Journal of Applied Physiology* 70: 399-404.

Pfeiffer, J.M., E.W. Askew, D.E. Roberts, S.M. Wood, J.E. Benson, S.C. Johnson, and M.S. Freedman. 1999. Effect of antioxidant supplementation on urine and blood markers of oxidative stress during extended moderate-altitude training. *Wilderness and Environmental Medicine* 10: 66-74.

Prefaut, C., F. Anselme-Poujol, and C. Caillaud. 1997. Inhibition of histamine release by nedocromil sodium reduces exercise-induced hypoxemia in master athletes. *Medicine and Science in Sports and Exercise* 29: 10-16.

Pyne, D.B. 1994. Exercise-induced muscle damage and inflammation: A review. *Australian Journal of Science and Medicine in Sport* 26: 49-58.

Reeves, J.T., E.E. Wolfel, H.J. Green, R.S. Mazzeo, A.J. Young, J.R. Sutton, and G.A. Brooks. 1992. Oxygen transport during exercise at high altitude and the lactate paradox: Lessons from Operation Everest II and Pikes Peak. In *Exercise and sport sciences reviews*, edited by K.B. Pandolf, pp. 275-296. New York: Macmillan.

Reeves, J.T., S. Zamudio, T.E. Dahms, I. Asmus, B. Braun, G.E. Butterfield, R.G. McCullough, S.R. Muza, P.B. Rock, and L.G. Moore. 2001. Erythropoiesis in women during 11 days at 4,300 m is not affected by menstrual cycle phase. *Journal of Applied Physiology* 91: 2579-2586.

Reif, D.W. 1992. Ferritin as a source of iron for oxidative damage. *Free Radical Biology and Medicine* 12: 417-427.

Rivera-Ch, M., A. Gamboa, F. Leon-Velarde, J.-A. Palacios, D.F. O'Connor, and P.A. Robbins. 2003. High-altitude natives living at sea level acclimatize to high altitude like sea-level natives. *Journal of Applied Physiology* 94: 1263-1268.

Roach, R.C., D. Maes, D. Sandoval, R.A. Robergs, M. Icenogle, H. Hinghofer-Szalkay, D. Lium, and J.A. Loeppky. 2000. Exercise exacerbates acute mountain sickness at simulated high altitude. *Journal of Applied Physiology* 88: 581-585.

Robergs, R.A., R. Quintana, D.L. Parker, and C.C. Frankel. 1998. Multiple variables explain the variability in the decrement in $\dot{V}O_2$max during acute hypobaric hypoxia. *Medicine and Science in Sports and Exercise* 30: 869-879.

Roberts, A.C., G.E. Butterfield, A. Cymerman, J.T. Reeves, E.E. Wolfel, and G.A. Brooks. 1996a. Acclimatization to 4300-m altitude decreases reliance on fat as a substrate. *Journal of Applied Physiology* 81: 1762-1771.

Roberts, A.C., J.T. Reeves, G.E. Butterfield, R.S. Mazzeo, J.R. Sutton, E.E. Wolfel, and G.A. Brooks. 1996b. Altitude and β-blockade augment glucose utilization during submaximal exercise. *Journal of Applied Physiology* 80: 605-615.

Roberts, A.D., P.J. Daley, D.T. Martin, A. Hahn, C.J. Gore, and R. Spence. 1998. Sea level $\dot{V}O_2$max fails to predict $\dot{V}O_2$max and performance at 1800m altitude. *Medicine and Science in Sports and Exercise* 30 (Suppl. 5): 111.

Roberts, D., and D.J. Smith. 1992. Training at moderate altitude: Iron status of elite male swimmers. *Journal of Laboratory and Clinical Medicine* 120: 387-391.

Roi, G.S., M. Giacometti, and S.P. Von Duvillard. 1999. Marathons at altitude. *Medicine and Science in Sports and Exercise* 31: 723-728.

Roncin, J.P., F. Schwartz, and P. D'Arbigny. 1996. EGb 761 in control of acute mountain sickness and vascular reactivity to cold exposure. *Aviation Space and Environmental Medicine* 67: 445-452.

Rose, M.S., C.S. Houston, C.S. Fulco, G. Coates, J.R. Sutton, and A. Cymerman. 1988. Operation Everest II: Nutrition and body composition. *Journal of Applied Physiology* 65: 2545-2551.

Roth, E. 1985. The sickle gene in evolution: A solitary wanderer or a nomad in a caravan of interacting genes. *Journal of the American Medical Association* 253: 2259-2260.

Rowbottom, D.G., D. Keast, and A. Morton. 1996. The emerging role of glutamine as an indicator of exercise stress and overtraining. *Sports Medicine* 21: 80-97.

Rusko, H.K., H. Kirvesniemi, L. Paavolainen, P. Vahasoyrinki, and K.P. Kyro. 1996. Effect of altitude training on sea level aerobic and anaerobic power of elite athletes. *Medicine and Science in Sports and Exercise* 28 (Suppl. 5): S124.

Saltin, B., H. Larsen, N. Terrados, J. Bangsbo, T. Bak, C.K. Kim, J. Svedenhag, and C.J. Rolf. 1995. Aerobic exercise capacity at sea level and altitude in Kenyan boys, junior and senior runners compared with Scandinavian runners. *Scandinavian Journal of Medicine and Science in Sports* 5: 209-221.

Sandoval, D.A., and K.S. Matt. 2002. Gender differences in the endocrine and metabolic responses to hypoxic exercise. *Journal of Applied Physiology* 92: 504-512.

Sandoval, D.A., and K.S. Matt. 2003. Effects of the oral contraceptive pill cycle on physiological responses to hypoxic exercise. *High Altitude Medicine and Biology* 4: 61-72.

Sato, H., J.W. Severinghaus, and P. Bickler. 1994. Time course of augmentation and depression of hypoxic ventilatory responses at altitude. *Journal of Applied Physiology* 77: 313-316.

Sato, H., J.W. Severinghaus, F.L. Powell, F.D. Xu, and M.J. Spellman. 1992. Augmented hypoxic ventilatory response in men at altitude. *Journal of Applied Physiology* 73: 101-107.

Sawka, M.N., V.A. Convertino, E.R. Eichner, S.M. Schneider, and A.J. Young. 2000. Blood volume: Importance and adaptation to exercise training, environmental stresses, and trauma/sickness. *Medicine and Science in Sports and Exercise* 32: 332-348.

Schouweiler, C.M., and J. Stray-Gundersen. 2002. Individual variation in the decrease of VO_2 at 1800 m in elite female cross-country skiers. *Medicine and Science in Sports and Exercise* 34 (Suppl. 5): S223.

Sevre, K., B. Bendz, E. Hanko, A.R. Nakstad, A. Hauge, J.I. Kasin, J.D. Lefrandt, A.J. Smit, I. Eide, and M. Rostrup. 2001. Reduced autonomic activity during stepwise exposure to high altitude. *Acta Physiologica Scandinavica* 173: 409-417.

Shave, R.E., G.P. Whyte, E. Dawson, K.P. George, D. Gaze, and P.O. Collinson. 2002. Evidence of secondary cardiac damage following prolonged exercise in hypoxia. *Medicine and Science in Sports and Exercise* 34 (Suppl. 5): S286.

Sheel, A.W., M.R. Edwards, G.S. Hunte, and D.C. McKenzie. 2001. Influence of inhaled nitric oxide on gas exchange during normoxic and hypoxic exercise in highly trained cyclists. *Journal of Applied Physiology* 90: 926-932.

Simon-Schnass, I. 1994. Risk of oxidative stress during exercise at high altitude. In *Exercise and oxygen toxicity*, edited by C.K. Sen, L. Packer, and O. Hanninen, pp. 191-210. Amsterdam: Elsevier.

Simon-Schnass, I., and H. Pabst. 1988. Influence of vitamin E on physical performance. *International Journal of Vitamin and Nutrition Research* 58: 49-54.

Squires, R.W., and E.R. Buskirk. 1982. Aerobic capacity during acute exposure to simulated altitude, 914 to 2286 meters. *Medicine and Science in Sports and Exercise* 14: 36-40.

Stenberg, J., B. Ekblom, and R. Messin. 1966. Hemodynamic response to work at simulated altitude, 4000 m. *Journal of Applied Physiology* 21: 1589-1594.

Stray-Gundersen, J., C. Alexander, A. Hochstein, D. deLemos, and B.D. Levine. 1992. Failure of red cell volume to increase to altitude exposure in iron deficient runners. *Medicine and Science in Sports and Exercise* 24 (Suppl.): S90.

Subudhi, A.W., S.L. Davis, R.W. Kipp, and E.W. Askew. 2001. Antioxidant status and oxidative stress in elite alpine ski racers. *International Journal of Sport Nutrition and Exercise Metabolism* 11: 32-41.

Svedenhag, J., B. Saltin, C. Johansson, and L. Kaijser. 1991. Aerobic and anaerobic exercise capacities of elite middle-distance runners after two weeks of training at moderate altitude. *Scandinavian Journal of Medicine and Science in Sports* 1: 205-214.

Sylvester, J.T., A. Cymerman, G. Gurtner, O. Hottenstein, M. Cote, and D. Wolfe. 1981. Components of alveolar-arterial O_2 gradient during rest and exercise at sea level and high altitude. *Journal of Applied Physiology* 50: 1129-1139.

Torre-Bueno, J.R., P.D. Wagner, H.A. Saltzman, G.E. Gale, and R.E. Moon. 1985. Diffusion limitation in normal humans during exercise at sea level and simulated altitude. *Journal of Applied Physiology* 58: 989-995.

Townsend, N.E., C.J. Gore, A.G. Hahn, M.J. McKenna, R.J. Aughey, S.A. Clark, T. Kinsman, J.A. Hawley, and C.-M. Chow. 2002. Living high-training low increases hypoxic ventilatory response in well-trained endurance athletes. *Journal of Applied Physiology* 93: 1498-1505.

Tsai, L., A. Pousette, K. Carlstrom, M. Askenberger, and C. Johansson. 1992. Anabolic and catabolic hormonal response of elite runners to training at high altitude. *Scandinavian Journal of Medicine and Science in Sports* 2: 10-15.

Uchakin, P., E. Gotovtseva, B. Levine, and J. Stray-Gundersen. 1995. Neuroimmunohumoral changes associated with altitude training. *Medicine and Science in Sports and Exercise* 27 (Suppl. 5): S174.

Vallier, J.M., P. Chateau, and C.Y. Guezennec. 1996. Effects of physical training in a hypobaric chamber on the physical performance of competitive triathletes. *European Journal of Applied Physiology* 73: 471-478.

Vasankari, T.J., U.M. Kujala, H. Rusko, S. Sarna, and M. Ahotupa. 1997. The effect of endurance exercise at moderate altitude on serum lipid peroxidation and antioxidant functions in humans. *European Journal of Applied Physiology* 75: 396-399.

Vasankari, T.J., H. Rusko, U.M. Kujala, and I.T. Huhtaniemi. 1993. The effect of ski training at altitude and racing on pituitary, adrenal and testicular function in men. *European Journal of Applied Physiology* 66: 221-225.

Vogel, J.A., and J.E. Hansen. 1967. Cardiovascular function during exercise at high altitude. In *The International Symposium on the Effects of Altitude on Physical Performance*. Chicago: The Athletic Institute.

Vogel, J.A., J.E. Hansen, and C.W. Harris. 1967. Cardiovascular responses in man during exhaustive work at sea level and high altitude. *Journal of Applied Physiology* 25: 531-539.

Vogel, J.A., L.H. Hartley, J.C. Cruz, and R.P. Hogan. 1974. Cardiac output during exercise in sea level residents at sea level and high altitude. *Journal of Applied Physiology* 36: 169-172.

Wagner, P.D., G.E. Gale, R.E. Moon, J.R. Torre-Bueno, B.W. Stolp, and H.A. Saltzman. 1986. Pulmonary gas exchange in humans exercising at sea level and simulated altitude. *Journal of Applied Physiology* 61: 260-270.

Welch, H.G. 1987. Effects of hypoxia and hyperoxia on human performance. *Exercise and Sport Sciences Reviews* 15: 191-220.

West, J.B. 1982. Diffusion at high altitude. *Federation Proceedings* 41: 2128-2130.

West, J.B. 2002a. Potential use of oxygen enrichment of room air in mountain resorts. *High Altitude Medicine and Biology* 3: 59-64.

West, J.B. 2002b. Commuting to high altitude: Value of oxygen enrichment of room air. *High Altitude Medicine and Biology* 3: 223-235.

Wetter, T.J., Z. Xiang, D.A. Sonetti, H.C. Haverkamp, A.J. Rice, A.A. Abbasi, K.C. Meyer, and J.A. Dempsey. 2002. Role of lung inflammatory mediators as a cause of exercise-induced arterial hypoxemia in young athletes. *Journal of Applied Physiology* 93: 116-126.

Weyand, P.G., C.S. Lee, R. Martinez-Ruiz, M.W. Bundle, M.J. Bellizzi, and S. Wright. 1999. High-speed running performance is largely unaffected by hypoxic reductions in aerobic power. *Journal of Applied Physiology* 86: 2059-2064.

White, D.P., K. Gleeson, C.K. Pickett, A.M. Rannels, A. Cymerman, and J.V. Weil. 1987. Altitude acclimatization: Influence on periodic breathing and chemoresponsiveness during sleep. *Journal of Applied Physiology* 63: 401-412.

Whyte, G.P., R.E. Shave, E. Dawson, K.P. George, D. Gaze, and P.O. Collinson. 2002. Left ventricular function following prolonged exercise in normobaric hypoxia. *Medicine and Science in Sports and Exercise* 34 (Suppl. 5): S286.

Wilber, R.L., S.D. Drake, J.L. Hesson, J.A. Nelson, J.T. Kearney, G.M. Dallam, and L.L. Williams. 2000. Effect of altitude training on serum creatine kinase activity and serum cortisol concentration in triathletes. *European Journal of Applied Physiology* 81: 140-147.

Wilber, R.L., P.L. Holm, D.M. Morris, G.M. Dallam, and S.D. Callan. 2003. Effect of F_iO_2 on physiological responses and cycling performance at moderate altitude. *Medicine and Science in Sports and Exercise* 35: 1153-1159.

Wolfel, E.E., M.A. Selland, R.S. Mazzeo, and J.T. Reeves. 1994. Systemic hypertension at 4300 m is related to sympathoadrenal activity. *Journal of Applied Physiology* 76: 1643-1650.

Wozniak, A., G. Drewa, G. Chesy, A. Rakowski, M. Rozwodowska, and D. Olszewska. 2001. Effect of altitude training on the peroxidation and antioxidant enzymes of sportsmen. *Medicine and Science in Sports and Exercise* 33: 1109-1113.

Young, A.J., W.J. Evans, A. Cymerman, K.B. Pandolf, J.J. Knapik, and J.T. Maher. 1982. Sparing effect of chronic high-altitude exposure on muscle glycogen utilization. *Journal of Applied Physiology* 52: 857-862.

Young, P.M., P.B. Rock, C.S. Fulco, L.A. Trad, V.A. Forte, and A. Cymerman. 1987. Altitude acclimatization attenuates plasma ammonia accumulation during submaximal exercise. *Journal of Applied Physiology* 63: 758-764.

Zielinski, J., M. Koziej, M. Mankowski, A.S. Sarybaev, J.S. Tursalieva, I.S. Sabirov, A.S. Karamuratov, and M.M. Mirrakhimov. 2000. The quality of sleep and periodic breathing in healthy subjects at an altitude of 3200 m. *High Altitude Medicine and Biology* 1: 331-336.

Part II

Altitude Training and Athletic Performance

Chapter 3
PERFORMANCE AT SEA LEVEL FOLLOWING ALTITUDE TRAINING

Several scientific studies have been conducted since the 1960s for the purpose of determining the effect of altitude training on athletic performance at sea level. To review all the scientific studies on this topic is beyond the scope of this chapter. Instead the aim is to summarize the relevant findings of several altitude training studies within the context of athletic performance at sea level. Figure 3.1 provides an overview of the chapter, which is organized according to two forms of altitude training: traditional "live high-train high" (LHTH) altitude training and contemporary "live high-train low" (LHTL) altitude training. For more detailed information on altitude training and athletic performance at sea level, the reader is referred to comprehensive review articles by Bailey and Davies (1997), Boning (1997), Fulco et al. (1998, 2000), Hahn (1991), Levine (2002), Wilber (2001), and Wolski et al. (1996). Chapter 6 ("Current Practices and Trends in Altitude Training") provides additional information on LHTL altitude training.

Factors Affecting the Results of Altitude Training Studies

Despite many years of research, we are still not able to say *conclusively* whether or not altitude training leads to improvements in sea level performance. The results of over three decades of altitude training research have simply not been consistent or reproducible. The discrepancy in these findings may result from several factors that one must take into consideration when evaluating the results of altitude training studies—particularly those that have failed to show a positive effect, that is, an improvement in sea level performance following altitude training.

One factor to be considered is the iron status of the athletes who participated in the investigations. As described in chapter 2, Stray-Gundersen et al. (1992) reported that altitude-induced erythropoiesis did not take place in endurance athletes who were diagnosed as iron deficient (serum ferritin < 20 ng · ml^{-1} for females; < 30 ng · ml^{-1}

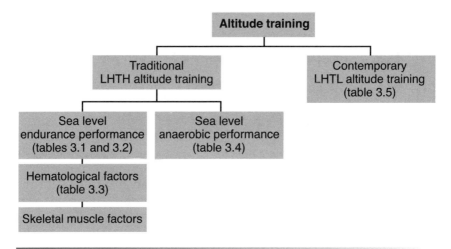

Figure 3.1 Organization of chapter 3 according to the scientific investigations that have examined the effects of two forms of altitude training on athletic performance at sea level: traditional "live high-train high" (LHTH) altitude training and contemporary "live high-train low" (LHTL) altitude training.

for males) prior to completion of a four-week altitude training camp at 2,500 m (8,200 ft). It is possible that some of the athletes involved in altitude training studies have had abnormally low ferritin levels before living and training at altitude, and thus may not have experienced an increase in red blood cell mass and hemoglobin. In turn, this may explain why some of those studies failed to demonstrate improvements in $\dot{V}O_2$max or endurance performance, or both, subsequent to altitude training.

The training stimulus at altitude is another factor that may have affected the results of scientific studies on altitude training and sea level endurance performance. For example, in one of the early altitude training studies conducted by Buskirk et al. (1967), the athletes were forced to markedly reduce their workout training intensity due to the hypoxic conditions of 4,000 m (13,120 ft). As a consequence of the decrease in training intensity, these athletes may have been *detrained* when they returned to sea level; and the detraining may have adversely affected $\dot{V}O_2$max and endurance performance. In addition, the training status of subjects may account for the inconsistent results reported in altitude training studies. Some of the subjects who have participated in altitude training studies have been elite, national team athletes, whereas others have been lesser-trained or recreational athletes.

Inconsistent results may also be attributed to differences in the hypoxic stimulus that the subjects were exposed to. By hypoxic stimulus we are referring to the elevation that the athletes lived and trained at, as well as the duration of altitude exposure. The hypoxic stimulus has varied greatly among altitude training studies.

Overtraining may also have confounded the performance results of some altitude training studies. For example, Vasankari et al. (1993) evaluated elite cross-country and biathlon skiers following a 14- to 18-day training and racing period at 1,650 m (5,410 ft); they reported a significant 29% increase in serum cortisol in the skiers following altitude training. As described in chapter 2, elevated serum cortisol has been associated with overtraining symptoms such as skeletal muscle catabolism (Hakkinen et al. 1989) and immunosuppression (Nieman et al. 1994, 1995a, 1995b; Nieman 1997). In addition, serum cortisol and other stress hormones may suppress erythropoiesis (Berglund 1992). Training-induced (or non-training-induced) injury or infection may also inhibit red blood cell (RBC) production via an increase in inflammatory cytokines (Fandrey and Jelkmann 1991). Thus, it is possible that some athletes failed to demonstrate an improvement in $\dot{V}O_2$max and performance because they returned to sea level in an overworked or overtrained state.

A final factor that we must consider in evaluating the performance results of altitude studies is the individual response of athletes to altitude training. Chapman et al. (1998) speculated that each athlete may need to follow an individualized altitude training program that allows for optimal physical training and physiological adaptation. In other words, one athlete may obtain the best physiological and performance results by following an LHTL altitude training program, whereas another athlete may get similar results by using traditional LHTH altitude training. Thus, it is possible that athletes who participated in some of the studies summarized in this chapter may not have been exposed to an altitude training regimen that allowed for the optimal development of physiological and performance-enhancing benefits. Individual responses may also be a factor when one is evaluating postaltitude performance. For example, Gore et al. (1998) reported that postaltitude cycling performance (4,000-m time trial) was very individual among elite track cyclists when evaluated on days 4, 9, and 21 postaltitude, suggesting that optimal performance for each athlete may occur at a different time point in the weeks following altitude training.

Traditional Live High-Train High Altitude Training

This part of the chapter focuses on the key findings of several scientific studies that have used traditional LHTH altitude training. The first section summarizes the important results regarding the effects of LHTH on sea level endurance performance. The next section reviews the effects of LHTH altitude training on hematological and skeletal muscle factors that influence endurance performance. A final section summarizes the key findings regarding the effects of LHTH on sea level anaerobic performance.

Sea Level Endurance Performance

Most of the scientific studies on altitude training have evaluated the effect of LHTH altitude training on sea level endurance performance. The initial studies were conducted in the mid-1960s. The LHTH method of altitude training has largely been replaced by "live high-train low" altitude training, which is described later in this chapter. In general, the results of several LHTH altitude training studies are equivocal; that is, some studies demonstrated that LHTH enhanced endurance performance upon return to sea level, whereas others showed that sea level endurance performance was not enhanced after LHTH altitude training.

LHTH Enhances Sea Level Endurance Performance: A Summary of the Scientific Literature

In general, the subjects who participated in these LHTH studies were either well-trained or elite aerobic athletes who lived and/or trained for 11 to 70 days at actual or simulated elevations ranging from 2,100 m to 4,000 m (6,890-13,120 ft). In each of these investigations, the athletes were evaluated within five days after completion of altitude training; in a few studies the athletes were evaluated at additional time points during a postaltitude period lasting two to three weeks. Performance measures in these studies included sport-specific time trials (run = 1 mile, 3 miles, 10,000 m; cycle = 4,000 m, 40 km); work capacity tests conducted on sport- and non-sport-specific ergometers; and endurance time during an exhaustive incremental exercise test.

One of the more notable studies in this group of LHTH investigations was conducted by Daniels and Oldridge (1970). Six male elite distance runners from the U.S. national team lived and trained for periods of 14, 14, 7, and 7 days at 2,300 m/7,544 ft (Alamosa, CO). The athletes returned to sea level for five days (SL 2) after the first 14-day altitude training block, and returned to sea level again for five days (SL 3) after the second 14-day altitude training block (figure 3.2). During these five-day periods at sea level, the runners competed in sanctioned national or international competitions. Most of the runners completed the study after the first seven-day altitude training block and were evaluated within five days (SL Post) after returning from altitude. "Normal sea level training" was conducted at altitude, and "hard training sessions" were included from the beginning of the altitude training camp. Compared with prealtitude values, $\dot{V}O_2$max increased by 4% at SL 2 (statistics not reported, n = 6), and by 5% when measured within five days after completion of all altitude training blocks. In addition, 3-mile race pace improved by 3% versus prealtitude values when measured within five days of completion of all altitude training blocks. Most notable was the fact that five of the six runners established 14 personal bests in sanctioned 1-mile and 3-mile races, which the athletes competed in during SL 2, SL 3, and SL Post (figure 3.2). One runner broke the world record for the 1 mile during SL 2 and set it again during SL 3.

Less than half of the LHTH investigations in this category evaluated changes in hemoglobin concentration; and among those studies, none showed a significant increase as a result of LHTH altitude training. Postaltitude improvements in $\dot{V}O_2$max varied from 1% to 10%, but only the 10% improvement found on the 16th day postaltitude in the study by Burtscher et al. (1996) was statistically significant. All of these LHTH studies, however, indicated that sea level endurance performance was enhanced following altitude training. Improvement in sea level endurance performance ranged from 4% to 33%. It should be emphasized that nearly 60% of the studies did not include a sea level control group in the experimental design, which makes their conclusions regarding performance debatable. In other words, for those studies that did not include a sea

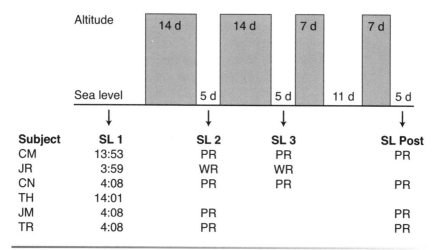

Figure 3.2 The experimental design of the altitude training study conducted by Daniels and Oldridge (1970) in which elite U.S. male distance runners completed intermittent two-week and one-week altitude training camps at 2,300 m/7,544 ft.

Adapted, by permission, from J. Daniels and N. Oldridge, 1970, "The effects of alternate exposure to altitude and sea level in world-class middle-distance runners," *Medicine and Science in Sports* 2: 108.

level control group the question remains, "Was the improvement in sea level endurance performance due to altitude training per se, or due to some other variable independent of altitude training that may have contributed to the enhancement of sea level performance?" Therefore, the investigations that did include a sea level control group (Asano et al. 1986; Burtscher et al. 1996; Terrados et al. 1988) should be considered the most credible among studies that produced results in support of altitude training for the enhancement of sea level endurance performance. Table 3.1 on pages 90-91 presents a summary of the scientific literature whose results support the use of LHTH altitude training for the enhancement of sea level endurance performance in trained aerobic athletes.

LHTH Does Not Enhance Sea Level Endurance Performance: A Summary of the Scientific Literature

In general, the subjects who participated in the LHTH studies reviewed here were either well-trained or elite aerobic athletes who lived and/or trained for 11 to 63 days at actual or simulated elevations ranging from 1,500 m to 4,000 m (4,920-13,120 ft). In most of these investigations, the athletes were evaluated within seven days after completion of altitude training; in a few studies the athletes were evaluated at additional time points during a postaltitude period lasting two to three weeks. Performance measures in these studies included sport-specific time trials (rowing = 2,500 m; run = 880 yd, 1,000 m, 1 mile, 2 miles, 3 miles, 5,000 m; swim = 100 m, 200 yd, 200 m, 500 yd); work capacity tests conducted on sport- and non-sport-specific ergometers; and endurance time during an incremental exhaustive exercise test.

One of the classic studies that did not demonstrate beneficial performance effects following LHTH altitude training was conducted by Adams et al. (1975). The subjects were 12 trained male distance runners who were divided into two training groups. One group (SL→ALT) trained at sea level while the other group (ALT→SL) trained at 2,300 m/7,544 ft (U.S. Air Force Academy, Colorado Springs, CO). The training program consisted of running 20 km (12.4 miles) per day at the same relative intensity (75% $\dot{V}O_2$max). Both groups trained at their respective locations for 20 days, then crossed over to the opposite training location for an additional 20 days. Postaltitude tests were conducted within one to three days after the runners returned from their respective 20-day altitude training blocks. Compared with baseline values, treadmill $\dot{V}O_2$max was lower in both the SL→ALT group (–1%, no statistically significant difference [NSD]) and the ALT→SL group (–4%, $p < 0.05$) after three weeks of altitude training (figure 3.3 on page 92). In terms of running performance, 2-mile run time was 7 sec faster (1%, NSD) in the SL→ALT runners, whereas the ALT→SL runners were 7 sec slower (–1%, NSD) following altitude training. The authors concluded that three weeks of living and training at moderate altitude had no effect on $\dot{V}O_2$max and running performance in trained distance runners. They further speculated that LHTH altitude training may have actually been detrimental to running performance and $\dot{V}O_2$max in many of the athletes they evaluated.

A few of the LHTH investigations in this category showed a significant improvement in hemoglobin concentration following altitude training; however, most indicated no improvement in sea level $\dot{V}O_2$max. In addition, all of these studies indicated that sea level endurance performance was not enhanced following LHTH altitude training, and a few showed that endurance performance was worse. Approximately 65% of these investigations included a sea level control group in the experimental design, thereby enhancing the credibility of their findings. Table 3.2 on pages 93-95 presents a summary of the scientific literature whose results do not support the use of LHTH altitude training for the enhancement of sea level endurance performance in trained aerobic athletes.

Hematological Factors

One of the primary reasons endurance athletes live and train at altitude is to stimulate an increase in erythropoietin (EPO), which in turn may produce an increase in RBC mass and hemoglobin concentration. As described in chapter 1, an increase in RBCs and hemoglobin will improve oxygen transport to the working muscles during submaximal and maximal exercise, contributing to an enhancement in endurance performance.

The scientific investigations reviewed in this section examined the effect of LHTH altitude training on hematological factors that influence endurance performance. The subjects who participated in these studies were either well-trained or elite aerobic athletes who lived and trained for 7 to 70 days at actual or simulated elevations ranging from 1,500 m to 4,100 m (4,920-13,450 ft). Criterion measures in these studies included serum EPO, reticulocytes, RBC mass, hemoglobin, and hematocrit. With few exceptions, measurements of serum EPO and reticulocytes were completed within the first five days after the subjects arrived at altitude; and measurements of RBC mass, hemoglobin, and hematocrit were made within the first week after return to sea level. Several of these investigations also evaluated the effect of altitude-induced hematological changes on sea level endurance performance.

Less than half of these studies evaluated the effect of LHTH altitude training on serum EPO; only two of those investigations reported a significant increase in serum EPO when measured within the first five days at altitude. Increments in serum EPO ranged from 31% to 54%. The same two studies reported that reticulocyte count was significantly

Table 3.1 Scientific Literature Supporting the Use of LHTH Altitude Training for the Enhancement of Sea Level Performance

Author	Altitude (m/ft)	Subjects	SL control	Altitude exposure (days)	Post-altitude SL test (day)	Δ Hb (%)	Δ $\dot{V}O_2$max (%)	Performance test	Δ Performance[a] (%)	Additional results/ comments
Mizuno et al. 1990	Live: 2,100/ 6,890 Train: 2,700/ 8,855	Male elite CC skiers (n = 10) Danish NT	No	14	2	NR	NSD	TM run time to exhaustion	17*	
Daniels and Oldridge 1970	2,300/ 7,544	Male elite runners (n = 6) U.S. NT	No	14/14/7[b]	1-5	NR	5*	1) 3-mile TT 2) 1- and 3-mile SC	1) 3* 2) 14 PRs, 1 WR	
Terrados et al. 1988	2,300/ 7,544[c]	Male elite cyclists (n = 8)	Yes	21-28	1-2	4 NSD	3 NSD	1) Work capacity (kJ) in exhaustive cycling test 2) Cycling maximal power output (W)	1) 33* 2) 12*	
Burtscher et al. 1996	2,315/ 7,595	Male recreational runners (n = 10)	Yes	12	3 16	NR NR	1 NSD 10*	PA 3: work capacity (kJ) in exhaustive cycling test PA 16: work capacity (kJ) in exhaustive cycling test	8* 16*	

Study	Altitude (m)	Subjects	Hypobaric chamber	Days at altitude	Days post-altitude tested			Performance test	Δ (%)	Notes
Gore et al. 1998	2,690/ 8,825	Male elite cyclists (n = 8) Australian NT	No	31	4, 9, 21	NSD	NSD	4,000-m TT	4*	4% Improvement based on the best time recorded by each athlete on any one of the three PA days
Dill and Adams 1971	3,090/ 10,135	Male champion HS runners (n = 6)	No	17	1	NR	4#	TM run time to exhaustion	24#	
Asano et al. 1986	4,000/ 13,120[c]	Male elite runners (n = 5)	Yes	70	"after"	NSD	NSD	10-km TT	5*	

Studies are listed by ascending altitude.

Δ = change.

[a] Positive value represents an improvement in postaltitude versus prealtitude sea level endurance performance.

[b] 14-day and 7-day training periods at altitude were separated by 5-day periods at sea level during which time subjects raced in sanctioned competitions.

[c] Hypobaric chamber.

* Significant difference versus prealtitude ($p < 0.05$).

Statistical analysis not reported.

CC = cross-country; Hb = hemoglobin (g · dL^{-1}); HS = high school; kJ = kilojoule; LHTH = "live high-train high"; NR = not reported/measured; NSD = no statistically significant difference; NT = national team; PA = postaltitude; PR = personal record; SC = sanctioned competition; SL = sea level; TM = treadmill; TT = time trial; $\dot{V}O_2$max = maximal oxygen consumption; W = watt; WR = world record.

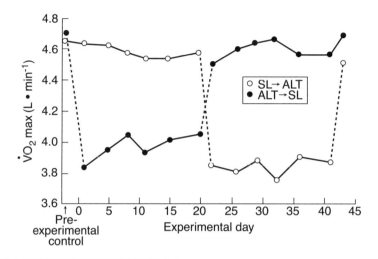

Figure 3.3 Schematic diagram showing the experimental crossover design and the maximal oxygen consumption ($\dot{V}O_2$max) results for two experimental groups at sea level (Davis, CA) and at altitude (U.S. Air Force Academy, Colorado Springs, CO 2,300 m/7,544 ft). Broken lines represent the change in site for the groups. SL → ALT = trained for 20 days at sea level, then trained for 20 days at altitude; ALT → SL = trained for 20 days at altitude, then trained for 20 days at sea level.

Reprinted, by permission, from W.C. Adams et al., 1975, "Effects of equivalent sea-level and altitude training on $\dot{V}O_2$max and running performance," *Journal of Applied Physiology* 39: 263.

higher (7-100%) during the first five days of altitude training. Several studies addressed the effect of LHTH altitude training on postaltitude hemoglobin concentration and hematocrit. Results of those investigations were somewhat equivocal, with approximately 60% showing that altitude training did not alter hemoglobin or hematocrit levels when measured during the week after return to sea level. Among those studies that demonstrated significant altitude-induced changes in hemoglobin or hematocrit, increments ranged from 5% to 9%. Only one investigation dealt with the effect of LHTH altitude training on RBC mass, showing a significant postaltitude increase of 10%.

Approximately 75% of the studies evaluating hematological factors that affect endurance performance also examined the effect of LHTH altitude training on postaltitude sea level endurance performance. Among those investigations, none demonstrated significant improvements in both hematological factors and sea level endurance performance. Approximately 40% indicated that hematological parameters were not affected by altitude training but that postaltitude endurance performance was enhanced, whereas 15% showed significant increments in

Table 3.2 Scientific Literature Not Supporting the Use of LHTH Altitude Training for the Enhancement of Sea Level Performance

Author	Altitude (m/ft)	Subjects	SL control	Altitude exposure (days)	Postaltitude SL test (day)	Δ Hb (%)	Δ $\dot{V}O_2$max (%)	Performance test	Δ Performance[a] (%)	Additional results/comments
Bailey et al. 1998	1,500/ 4,920	Male elite runners (n = 14) British NT	Yes	28	20	4 NSD	NR	1,000-m TT	NSD	LT and running economy NSD
Bailey et al. 1998	1,640/ 5,380	Male elite runners (n = 10) British NT	Yes	28	10 20	9 NSD 5 NSD	1 NSD -1 NSD	1,000-m TT 1,000-m TT	NR -2*	
Telford et al. 1996	1,760/ 5,775	Male elite runners (n = 9) Australian NT	Yes	28	1-7	3 NSD	3 NSD	3,200-m TT	NSD	3,200-m TT 10 sec faster after AT, but identical to improvement in SL control group
Jensen et al. 1993	1,822/ 5,975	Male elite rowers (n = 9) Italian NT	Yes	21	"within 7"	NR	NSD	6-min rowing ergometer test	NSD	
Chung et al. 1995	1,890/ 6,200	Female and male elite swimmers (n = 10) South Korean NT	Yes	21	7	F = 10* M = 4*	F = 0 M = 5*	Swimming NC; 100- and 200-m events	0.1-0.7*	Swimming NC held 6 weeks postaltitude
Ingjer and Myhre 1992	1,900/ 6,230	Male elite CC skiers (n = 7) Norwegian NT	No	21	1 14	5* 1	1 NSD 1 NSD	NR NR	NR NR	Significantly lower blood lactate during 6-min TM run at 90% $\dot{V}O_2$max on postaltitude day 1, but not postaltitude day 14. *(continued)*

Table 3.2 (continued)

Author	Altitude (m/ft)	Subjects	SL control	Altitude exposure (days)	Post-altitude SL test (day)	Δ Hb (%)	Δ $\dot{V}O_2$max (%)	Performance test	Δ Performance[a] (%)	Additional results/comments
Svedenhag et al. 1991	2,000/ 6,560	Male elite runners (n = 5) Swedish NT	yes	14	6 12	2 NSD 1 NSD	NSD NSD	TM run time to exhaustion TM run time to exhaustion	1 NSD −5 NSD	
Faulkner et al. 1967	2,300/ 7,544	Male collegiate swimmers (n = 15)	No	14	1	1 NSD	1 NSD	1) 200-yd TT 2) 500-yd TT	1) NSD 2) 1 NSD	
Faulkner et al. 1968	2,300/ 7,544	Male collegiate runners (n = 5)	No	42	3-6	NR	2*	1) 1-mile TT 2) 2-mile TT 3) 3-mile TT	1) 0-2* 2) −1-1* 3) 2-3*	
Adams et al. 1975[b]	2,300/ 7,544	1) Male trained runners (n = 6) 2) Male trained runners (n = 6)	Yes Yes	SL 20 → ALT 20 ALT 20 → SL 20	1-3 1-3	NR NR	−1 NSD −4*	2-mile TT 2-mile TT	1 NSD −1 NSD	SL → ALT group was 7 sec faster (9:03 vs. 9:10) ALT → SL group was 7 sec slower (9:22 vs. 9:15)
Levine and Stray-Gundersen 1997	2,500/ 8,200	Female and male collegiate runners (n = 13)	Yes	28	3	9*	5*	5-km TT	NSD	5-km TT slower by 3.3 sec 5-km TT on SL days 7, 14, 21 NSD versus SL day 3
Rahkila and Rusko 1982	2,600/ 8,530	Male trained CC skiers (n = 6)	Yes	11	"after"	5*	−3 NSD			

Study	Altitude	Subjects	Hypoxic chamber	Days at altitude	↑ (Hb, VO$_2$max) 6 (performance)	Δ Hb	Δ VO$_2$max	Performance test	Δ Performance	Comments
Hahn et al. 1992	3,100/ 10,170[c]	Female and male elite rowers (n = 8) Australian NT	Yes	19		−1 NSD	NSD	2,500-m rowing ergometer TT	1 NSD	
Buskirk et al. 1967	4,000/ 13,120	Male collegiate runners (n = 6)	No	63	3-7 10-15 3-7 10-15 3-7 10-15	NR	0 0 0 0 0	880-yd TT 880-yd TT 1-mile TT 1-mile TT 2-mile TT 2-mile TT	0 −6−−4* −6−−4* −6−−3* −4−0* −8−−4*	No improvement in performance for any athlete in any TT
Vallier et al. 1996	4,000/ 13,120[d]	Female and male elite triathletes (n = 5) French NT	No	21	7	−3 NSD	2 NSD	Maximal power output (W) during cycling exercise	NSD	

Studies are listed by ascending altitude.

Δ = change.
[a] Positive value represents an improvement in postaltitude versus prealtitude sea level endurance performance; negative value represents a poorer postaltitude versus prealtitude sea level endurance performance.
[b] Refer to text for explanation of study design.
[c] Hypoxic gas training (15.2% O$_2$).
[d] Hypobaric chamber.
* Significant difference versus prealtitude ($p < 0.05$).
# Statistical analysis not reported.

ALT = altitude; AT = altitude training; CC = cross-country; F = female; Hb = hemoglobin (g · dL^{-1}); LHTH = "live high-train high"; LT = lactate threshold; M = male; NC = national championships; NR = not reported/measured; NSD = no statistically significant difference; NT = national team; SL = sea level; TM = treadmill; TT = time trial; VO$_2$max = maximal oxygen consumption; W = watt.

serum EPO, hemoglobin, and/or hematocrit but no improvement in sea level endurance performance. Approximately 45% indicated that neither hematological parameters nor postaltitude endurance performance were affected by LHTH altitude training.

It should be noted, however, that several factors may have affected the hematological results of some of these LHTH studies. One factor may have been the iron status of the athletes who participated in the investigations. As previously discussed, athletes who have abnormally low ferritin levels prior to training at altitude may not realize an increase in RBC mass and hemoglobin. Since most of the studies reviewed in this section did not assess prealtitude iron status, it is difficult to determine whether iron deficiency affected the hematological results. It is possible that some of the athletes may have been moderately or severely iron deficient prior to LHTH altitude training.

Discrepancies in hematological results also may have been due to methodological differences in the measurement of total hemoglobin and RBC mass. Some investigations using the Evans blue dye technique showed significant increases in total hemoglobin following altitude exposure, whereas authors of studies using the carbon monoxide rebreathing technique failed to detect any significant changes in total hemoglobin as a result of altitude exposure. Ashenden et al. (1999) suggested that the Evans blue dye technique may measure a spurious increase in total hemoglobin and red cell mass because of increased albumin leakage from the vasculature due to altitude exposure.

Another factor that may have affected the hematological results is the plasma volume status of the athletes. Shifts in plasma volume are known to occur when athletes travel to altitude and again when they return to sea level. It is well known that changes in plasma volume can affect the measurement of hemoglobin and hematocrit. Thus, if changes in plasma volume are not controlled for, hemoglobin and hematocrit results must be interpreted with caution.

An additional factor that one must take into account in evaluating the hematological results is the individual response of athletes at altitude (Chapman et al. 1998). It is possible that some of the athletes who participated in these LHTH studies may not have been exposed to an altitude high enough to stimulate an increase in EPO and consequently affect the production of RBCs and hemoglobin.

Finally, it may be that some of these LHTH studies did not demonstrate significant improvements in hematological parameters because hemoglobin and hematocrit levels of elite aerobic athletes tend to be relatively high prior to altitude training. This may be a result of inherent genetic predisposition, the cumulative effect of many years of endurance training, or both. Nevertheless, it may be unrealistic to expect statistically significant increments in hemoglobin and hematocrit in elite athletes whose normal levels may be relatively high independent of altitude training (Gore et al. 1998). Table 3.3 summarizes the key findings of scientific research on the effects of LHTH on hematological variables.

Skeletal Muscle Factors

Although the primary reason endurance athletes train at altitude is to increase their RBC mass and hemoglobin concentration, there may be secondary physiological benefits to be gained as a result of altitude training. One of these potential changes is an increase in the number of skeletal muscle capillaries, which serve to enhance the rate of oxygen extraction from the blood. Another potential adaptation is an increase in mitochondrial oxidative enzyme activity, which may allow for greater utilization of oxygen for energy production. Scientific data in support of these potential altitude-induced physiological adaptations are minimal and equivocal, especially among trained and elite athletes.

Skeletal Muscle Capillarity

Capillaries are single-cell structures whose primary physiological function is the exchange of oxygen, carbon dioxide, nutrients, and metabolic waste products between the blood and tissues. In theory, an increase in the number of capillaries per unit area (capillaries per mm^2) or the number of capillaries surrounding each muscle fiber (capillaries per fiber) enhances the exercising muscles' ability to extract oxygen from the blood. Scientific studies pertinent to the effect of LHTH altitude training on skeletal muscle capillarization in trained or elite athletes are limited and equivocal. The following subsections summarize these studies according to ascending altitude. If a summary does not include mention of a sea level control group, the reader can assume that one was not included in the research design.

Saltin et al. (1995) Six female and male elite Scandinavian runners lived and trained for 14 days at 2,000 m/6,560 ft (Eldoret region of Kenya). Details of the training program at altitude were not reported. A group of six female and male runners trained during the same time at sea level. Postaltitude samples were extracted via needle biopsy from the vastus lateralis muscle within one week of completion of training. Following the two-week training period, there was no change in the number of capillaries per mm^2 or capillaries per fiber for either the altitude group or the sea level group compared with their respective baseline values. Thus, these data suggested that altitude training for 14 days at 2,000 m (6,560 ft) had no effect on the skeletal muscle capillarity of elite distance runners compared to similar training at sea level. However, it has been argued that a two-week altitude training camp,

Table 3.3 Scientific Literature About the Effect of LHTH Altitude Training on Hematological Factors

Author	Altitude (m/ft)	Subjects	SL control	Altitude exposure (days)	Δ EPO[a] (%)	Δ Reticulocytes[a] (%)	Δ RCM[b] (%)	Δ Hb[b] (%)	Δ Hct[b] (%)	Δ $\dot{V}O_2$max (%)	Postaltitude SL performance[b]
Bailey et al. 1998	1,500/ 4,920	Male elite runners (n = 14) British NT	Yes	28	NR	NR	NR	4 NSD	NR	NR	1,000-m TT not significantly faster
Bailey et al. 1998	1,640/ 5,380	Male elite runners (n = 10) British NT	Yes	28	NR	NR	NR	9 NSD	NR	1 NSD	1,000-m TT slower by 2%*
Klausen et al. 1991	Live: 1,700/ 5,575 Train: 2,700/ 8,855	Male trained CC skiers (n = 6)	No	7	31*	100*	NR	5*	NR	NR	NR
Telford et al. 1996	1,760/ 5,775	Male elite runners (n = 9) Australian NT	Yes	28	NR	NR	NR	3 NSD	NR	3 NSD	3,200-m TT faster by 10 sec, but identical to improvement in SL control group
Friedmann et al. 1999	1,800/ 5,905	Male boxers (n = 16) German NT	No	18	I = 54* P = 28*	I = 6.5* P = 7*	NR	I = -9* P = NSD	I = 3* P = -2*	NSD	TM $\dot{V}O_2$max and MSS run velocity not significantly improved
Chung et al. 1995	1,890/ 6,200	Female and male elite swimmers (n = 10) South Korean NT	Yes	21	NR	NR	NR	F = 10* M = 4*	F = 6* M = 4*	F = 0 M = 5*	Swimming performance improved in all altitude-trained swimmers at NC held 6 weeks postaltitude

Ingjer and Myhre 1992	1,900/ 6,230	Male elite CC skiers (n = 7) Norwegian NT	No	21	NR	NR	NR	5*	5*	1 NSD	NR
Svedenhag et al. 1997	1,900/ 6,230	Female and male elite CC skiers (n = 7) Swedish NT	No	30	NR	NR	NR	Increased*	Increased#	NR	NR
Svedenhag et al. 1991	2,000/ 6,560	Male elite runners (n = 5) Swedish NT	Yes	14	NR	NR	NR	2 NSD	NR	NSD	TM run time to exhaustion not significantly longer
Faulkner et al. 1967	2,300/ 7,544	Male collegiate swimmers (n = 15)	No	14	NR	NR	NR	1 NSD	2 NSD	1 NSD	200-yd and 500-yd TT not significantly faster
Terrados et al. 1988	2,300/ 7,544[c]	Male elite cyclists (n = 8)	Yes	21-28	NR	NR	NR	4 NSD	5 NSD	3 NSD	Cycling work capacity (kJ) improved by 33%*
Levine and Stray-Gundersen 1997	2,500/ 8,200	Female and male collegiate runners (n = 13)	Yes	28	NR	NR	10*	9*	NR	5*	5-km TT not significantly faster
Berglund et al. 1992	2,600/ 8,530	Female and male elite CC skiers (n = 20)	No	14	17 NSD	NR	NR	NR	NR	NR	NR

(continued)

Table 3.3 (continued)

Author	Altitude (m/ft)	Subjects	SL control	Altitude exposure (days)	Δ EPO[a] (%)	Δ Reticu-locytes[a] (%)	Δ RCM[b] (%)	Δ Hb[b] (%)	Δ Hct[b] (%)	Δ V̇O₂max (%)	Postaltitude SL performance[b]
Gore et al. 1998	2,690/ 8,825	Male elite cyclists (n = 8) Australian NT	No	31	NR	NR	NR	NSD	NR	NSD	4,000-m TT faster in 7 of 8 cyclists*
Hahn et al. 1992	3,100/ 10,170[d]	Female and male elite rowers (n = 8) Australian NT	Yes	19	NR	NR	NR	−1 NSD	−2 NSD	NSD	2,500-m rowing ergometer TT not significantly faster
Asano et al. 1986	4,100/ 13,450[c]	Male elite runners (n = 5)	Yes	70	NR	NR	NR	NSD	NSD	NSD	10-km TT faster by 5%*
Vallier et al. 1996	4,100/ 13,450[c]	Female and male elite triathletes (n = 5) French NT	No	21	28 NSD	53 NSD	NR	−3 NSD	1 NSD	1 NSD	Cycling maximal power output (W) not significantly improved

Studies are listed by ascending altitude.

Δ = change.
[a] Measured within 1 to 7 days after arriving at altitude. Exception: Berglund et al. (1992) on day 14.
[b] Measured within 1 to 7 days postaltitude. Exceptions: Gore et al. (1998) on days 4, 9, and 21; Bailey et al. (1998) on day 20.
[c] Hypobaric chamber.
[d] Supplemental hypoxic training (15.2% O_2).
* Significant difference versus prealtitude ($p < 0.05$).
# Statistical analysis not reported.

CC = cross-country; EPO = erythropoietin (mU · ml⁻¹); F = female; Hb = hemoglobin (g · dL⁻¹); Hct = hematocrit (%); I = iron-supplemented athletes; kJ = kilojoule; LHTH = "live high-train high"; M = male; MSS = maximal steady state; NC = national championships; NR = not reported/measured; NSD = no statistically significant difference; NT = national team; P = placebo athletes; RCM = red cell mass (ml · kg⁻¹); SL = sea level; TM = treadmill; TT = time trial; V̇O₂max = maximal oxygen consumption; W = watt.

such as the one used in this study, may not be long enough to induce favorable physiological changes.

Mizuno et al. (1990)   The subjects in the study by Mizuno et al. (1990) were 10 male elite cross-country skiers from the Danish national team. The athletes lived at 2,100 m (6,890 ft) and trained at 2,700 m (8,855 ft) for two weeks; specific details regarding the duration and intensity of the individual workouts were not reported. Postaltitude samples were extracted via needle biopsy from the gastrocnemius and triceps brachii muscles within 36 hr of completion of the training period. For the triceps brachii muscle, there was a significant increase in the number of capillaries per fiber (15%) following altitude training. Significant changes in capillarity were not observed in the gastrocnemius muscle. It was concluded that a 14-day training camp at moderate altitude may lead to improvement in skeletal muscle capillarity in elite cross-country skiers. However, because this study lacked a sea level control group, it is difficult to draw definitive conclusions regarding the effect of altitude training on skeletal muscle capillarity.

Terrados et al. (1988)   Eight male elite cyclists served as subjects in this study and were randomly assigned to two training groups. One group (ALT) trained in a hypobaric chamber at a simulated altitude of 2,300 m (7,544 ft), whereas the other group (SL) trained in a laboratory maintained at sea level environmental conditions. Both groups trained for 21 to 28 days, four to five days per week. Training consisted of 60 to 90 min of continuous cycle ergometer exercise at the same relative intensity (65-70% Wmax) followed by 45 to 60 min of interval training. Postaltitude samples were extracted via needle biopsy from the vastus lateralis muscle within one to two days of completion of the training period. Compared with prealtitude values, the ALT group showed improvements in the number of capillaries per mm^2 (36%) and capillaries per fiber (15%); however, these increments were not statistically different from prealtitude values. The SL group demonstrated a 21% increase (NSD) in the number of capillaries per mm^2 and an 8% decrease (NSD) in the number of capillaries per fiber. These data suggested that altitude training for 21 to 28 days at 2,300 m (7,544 ft) produced marked but statistically nonsignificant improvements in the number of capillaries per muscle fiber compared with equivalent training at sea level. It is possible that the increase in skeletal muscle capillarity reported in this study may have been due to a decrease in skeletal muscle fiber size, not to increased capillarization per se.

Skeletal Muscle Enzymes

Skeletal muscle enzymes are analyzed and quantified via needle biopsy techniques. Through use of this method, muscle samples can be analyzed for marker enzymes that reflect the activity of specific metabolic pathways. The following is a list of some specific metabolic pathways and their corresponding marker enzymes.

Metabolic Pathway	Enzyme
Glycogenolysis	Phosphorylase
Glycolysis	Phosphofructokinase (PFK)
	Lactate dehydrogenase (LDH)
Mitochondrial oxidation	
• Krebs cycle	Citrate synthase (CS)
	Succinate dehydrogenase (SDH)
• Beta-oxidation	Beta-hydroxyacyl-coA dehydrogenase (HADH)

A limited number of studies have addressed the effect of LHTH altitude training on the skeletal muscle enzyme activity of trained or elite athletes. The authors of these studies were unable to determine whether or not altitude training induces favorable changes in skeletal muscle enzymes that influence endurance performance. In the following summaries, which are presented by ascending altitude, if a sea level control group is not mentioned, the reader can assume that the research design did not include one.

Kuno et al. (1994) The subjects for the study by Kuno et al. (1994) were four male elite Nordic combined skiers from the Japanese national team. The athletes trained twice a day for four consecutive days in a hypobaric chamber at a simulated altitude of 2,000 m (6,560 ft). They performed workouts on either a cycle ergometer or a treadmill for 60 min at a heart rate intensity of 130 to 140 bpm. Phosphorus-31 nuclear magnetic resonance (^{31}P NMR) imaging was used to evaluate adenosine triphosphate (ATP) synthesis during exercise recovery and thus served as an indirect measure of mitochondrial oxidative metabolism. The athletes were evaluated during and after an exercise bout that required them to lie in a supine position and repeatedly lift the right leg in a fully extended position at a rate of 50 lifts per minute for 8 min. A 1-kg weight was attached to the right ankle during the exercise session. Compared with pretraining values, ATP resynthesis following the exercise bout was significantly enhanced by 19% following the four-day simulated altitude training period. The authors concluded that the observed improvement in ATP formation rate may have been due to increased mitochondrial oxidative capacity resulting from hypoxic training. However, these results must be interpreted with caution because a control group was not included in the research design. In addition, the criterion exercise test (repetitive leg lifts) may not have accurately evaluated the effect of the four-day hypoxic training period on mitochondrial enzyme activity.

Saltin et al. (1995) Details regarding the subjects and design of this study were described previously in the section on skeletal muscle capillarity. Postaltitude samples were extracted via needle biopsy from the vastus lateralis and gastrocnemius muscles within one week of completion of the training period conducted at 2,000 m (6,560 ft). For both the vastus lateralis and gastrocnemius, there were no changes in CS or HADH for either the altitude group or the sea level group compared with their respective baseline values. However, there was a significant decrease in LDH (−15%) in the gastrocnemius muscle of the altitude-trained group, suggesting a reduction in glycolytic enzyme activity. This decrease in glycolytic enzyme activity may have been related to a reduction in training intensity at 2,000 m (6,560 ft). However, since the training program was not described, it is difficult to draw this conclusion. In general, these data suggested that altitude training for 14 days at 2,000 m had no effect on mitochondrial oxidative enzymes compared with similar training at sea level, but that it may lead to a reduction in glycolytic enzyme activity.

Mizuno et al. (1990) Details of the subjects and design of this study were provided earlier in the discussion of skeletal muscle capillarity. Postaltitude samples were extracted via needle biopsy from the gastrocnemius and triceps brachii muscles within 36 hr of completion of the training period conducted at 2,100 m to 2,700 m (6,890-8,855 ft). For the gastrocnemius muscle, there was a significant reduction in the enzyme activity of CS (−13%), HADH (−10%), and phosphorylase (decrement not reported), whereas LDH was unaltered. A significant change in glycolytic or oxidative enzyme activity was not observed in the triceps brachii muscle following the two-week altitude training period. The decrease in oxidative enzyme activity observed in the gastrocnemius muscle was attributed to the fact that fibers in the gastrocnemius were used to a lesser extent during the cross-country "skating" technique than were fibers in the triceps brachii muscle; thus the gastrocnemius muscle may have been detrained during the altitude training block.

Terrados et al. (1988) Details regarding the subjects and design of this study have already been described in the section "Skeletal Muscle Capillarity" (page 97). Postaltitude samples were extracted via needle biopsy from the vastus lateralis muscle within one to two days of completion of the training period conducted at 2,300 m (7,544 ft). Following the training period, the ALT group showed decrements in PFK (−18%) and LDH (−25%) suggesting a reduction in glycolytic enzyme activity; these decrements were significantly lower compared with values in the SL group, which demonstrated an increase in the concentrations of both PFK and LDH. No significant changes were observed in CS or HADH for either the ALT or SL cyclists. Thus, these data suggested that altitude training for 21 to 28 days at 2,300 m (7,544 ft) resulted in decrements in glycolytic enzyme activity and had no effect on oxidative enzyme activity compared with similar training at sea level.

Summary

A limited number of studies have evaluated the effect of LHTH altitude training on skeletal muscle capillarity and enzyme activity in trained or elite endurance athletes. The subjects recruited for these studies lived and/or trained for 4 to 28 days at actual or simulated elevations ranging from 2,000 m to 2,700 m (6,560-8,855 ft). In most of these investigations, the athletes were evaluated within one to two days after completion of the altitude training period. Data from a limited number of studies are equivocal and do not provide definitive evidence that skeletal muscle capillarity is enhanced in elite aerobic athletes as a result of training at moderate altitude. In addition, altitude training does not appear to increase mitochondrial oxidative enzyme activity (CS, HADH) in elite endurance athletes, but may lead to reductions in glycolytic enzyme activity (PFK, LDH). This decrease in glycolytic enzyme activity may be due in part to the reduction in training intensity that frequently occurs with LHTH altitude training.

Sea Level Anaerobic Performance

Anaerobic performance is defined in this section as maximal or supra-maximal exercise lasting approximately 60 sec or less. Few scientific studies have addressed the effects of altitude training on anaerobic performance. The reason is that the scientific rationale supporting the use of altitude training for anaerobic performance is less compelling than the argument supporting the use of altitude training for the enhancement of endurance performance. For example, one of the purported benefits of altitude training is an increase in RBC mass and hemoglobin concentration. In turn, these hematological adaptations have the potential to enhance oxygen transport capacity and endurance performance upon return to sea level. In anaerobic events lasting approximately 60 sec or less, the contribution of aerobic metabolism to total energy production is minor. Thus, altitude-induced increments in RBC mass and hemoglobin are physiological adaptations that probably do not affect sea level anaerobic performance to a significant degree.

The subjects who participated in studies of the effects of LHTH on sea level anaerobic performance lived and trained for 11 to 28 days at actual or simulated elevations ranging from 1,600 m to 2,600 m (5,250-8,530 ft). Performance measures in these studies included sport-specific time trials (swim = 100 yd), velocity and power measured during a maximal anaerobic run test, and a supramaximal work capacity test conducted on a non-sport-specific ergometer. In addition, a few studies evaluated the effect of altitude training on oxygen deficit.

In terms of performance, the collective results of the studies reviewed do not support the use of LHTH altitude training for the enhancement of sea level anaerobic performance. One of the limiting factors in several of these studies was the fact that the subjects were endurance athletes whose training at altitude consisted primarily of aerobic workouts. This was the case because the primary purpose of most of these investigations was to evaluate the effect of LHTH altitude training on sea level endurance performance, whereas a secondary purpose was the assessment of sea level anaerobic performance.

One important physiological adaptation that may occur as a result of LHTH altitude training is an improvement in the capacity of the skeletal muscle to buffer the concentration of H^+. An enhanced H^+-buffering capacity could, in theory, have a beneficial effect on anaerobic performance. Mizuno et al. (1990) reported that relative to prealtitude values, there was a 29% increase ($p < 0.05$) in maximal oxygen deficit, as well as a 6% increase ($p < 0.05$) in the buffering capacity of the gastrocnemius muscle, in Danish national team cross-country skiers who lived at 2,100 m (6,890 ft) and trained at 2,700 m (8,855 ft) for 14 days. However, because the study lacked a sea level control group, it is difficult to draw definitive conclusions regarding this finding.

Anecdotal evidence and performance results suggest that anaerobic performance is enhanced during acute exposure to altitude. It appears that because of the decrease in air density, athletes are "faster" at moderate altitude in track sprint events lasting up to approximately 50 sec (Peronnet et al. 1991). However, the question remains whether long-term altitude training, that is, several consecutive days of sprinting "faster" at altitude, leads to an improvement in sprint performance upon return to sea level.

Table 3.4 presents a summary of the scientific literature pertinent to the use of LHTH altitude training for the enhancement of sea level anaerobic performance in trained athletes.

Contemporary Live High-Train Low Altitude Training

Since the mid-1990s, the LHTL method of altitude training has become a popular alternative to traditional LHTH altitude training. Athletes who use LHTL live and/or sleep at moderate altitude (2,000-2,700 m/

Table 3.4 Scientific Literature About the Use of LHTH Altitude Training for the Enhancement of Sea Level Anaerobic Performance

Author	Altitude (m/ft)	Subjects	SL control	Altitude exposure (days)	Postaltitude SL test (day)	Performance test	Δ Performance[a] (%)	Additional results/comments
Rusko et al. 1996	1,600-1,800/ 5,250-5,905	Male elite CC and biathlon skiers (n = 14) Finnish NT	Yes	18-28	<7	Anaerobic power during MART: 1) P_{5mM} 2) P_{max}	1) −5* 2) −2*	NSD in $\dot{V}O_2$max
Svedenhag et al. 1991	2,000/6,560	Male elite distance runners (n = 5) Swedish NT	Yes	14	6 12	Maximal O_2 deficit Maximal O_2 deficit	17* 9 NSD	
Mizuno et al. 1990	Live: 2,100/6,890 Train: 2,700/8,855	Male elite CC skiers (n = 10) Danish NT	No	14	2	Maximal O_2 deficit	29*	Buffering capacity of the triceps brachii and gastrocnemius muscles increased by 6%*
Faulkner et al. 1967	2,300/7,544	Male collegiate swimmers (n = 15)	No	14	1	100-yd TT	NSD	
Levine and Stray-Gundersen 1997	Live: 2,500/8,200 Train: 2,500/8,200	Female and male collegiate runners (n = 13)	Yes	28	3	Accumulated O_2 deficit	−9 NSD	
Rahkila and Rusko 1982	2,600/8,530	Male trained CC skiers (n = 6)	Yes	11	"after"	1-min maximal cycle ergometer test (kJ · kg⁻¹)	NSD	

Anaerobic performance is defined as maximal or supramaximal exercise lasting for approximately 1 min or less. Studies are listed by ascending altitude. Δ = change.

[a] Positive value represents an improvement in postaltitude versus prealtitude sea level anaerobic performance; negative value represents a poorer postaltitude versus prealtitude sea level anaerobic performance.

* Significant difference versus prealtitude (p < 0.05).

CC = cross-country; kJ = kilojoule; LHTH = "live high-train high"; MART = maximal anaerobic run test; NSD = no statistically significant difference; NT = ... = maximal power; SL = sea level; TT = time trial; $\dot{V}O_2$max = maximal oxygen consumption.

6,560-8,855 ft) while simultaneously training at low elevation (≤1,000 m/3,280 ft).

The initial study on LHTL altitude training was conducted by Levine and Stray-Gundersen (1997). The subjects for that investigation were 13 female and male collegiate runners. Following a four-week baseline period at sea level, the runners completed a 28-day training period in which they lived at 2,500 m/8,200 ft (Park City, UT) and trained at 1,250 m/4,100 ft (Salt Lake City, UT) (figure 3.4). Training consisted of alternate workouts of base training and interval training. Thirteen female and male collegiate runners, serving as a control group, followed the same training program at sea level. Compared with prealtitude values, postaltitude tests conducted on the third day following altitude training indicated significant improvements in the LHTL group for RBC mass (5%), hemoglobin (9%), and treadmill $\dot{V}O_2$max (5%). No improvements in RBC mass, hemoglobin, or $\dot{V}O_2$max were observed in the control group following the 28-day training period. In addition, an average 1% improvement ($p < 0.05$) in postaltitude 5,000-m run time was seen in the LHTL group (figure 3.5), an improvement that was equivalent to 13.4 sec. Performance in the 5,000-m run for the LHTL runners was similar on days 7, 14, and 21 postaltitude compared with day 3 postaltitude, suggesting that the beneficial effects of LHTL altitude training on running performance may last for up to three weeks postaltitude. In contrast, the control group did not show an improvement in 5,000-m run performance at any time following the 28-day training period. Improvement in 5,000-m run performance in the LHTL runners was also found to be moderately correlated ($r = 0.65$, $p < 0.01$) to a significant 5% improvement in treadmill $\dot{V}O_2$max.

Collectively, these data suggested that living at a relatively high altitude (2,500 m/8,200 ft) brought about significant increases in RBC mass and hemoglobin concentration. Simultaneous training at a lower elevation (1,250 m/4,100 ft) allowed these athletes to achieve running velocities similar to their sea level running velocities, purportedly inducing beneficial peripheral and neuromuscular adaptations. When the runners returned to sea level, significant improvements in $\dot{V}O_2$max and endurance performance were demonstrated—effects that were attributed to the hematological and neuromuscular adaptations resulting from four weeks of LHTL altitude training (Levine and Stray-Gundersen 1997).

In a modification of LHTL altitude training, the athlete lives and completes low-intensity base training at a relatively high elevation but does high-intensity interval workouts at low altitude. Stray-Gundersen et al. (2001) recently evaluated this variation of LHTL altitude training, referred to as "live high-base train high-interval train low" (HiHiLo). Twenty-two female and male elite distance runners served as subjects.

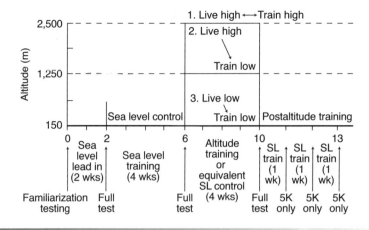

Figure 3.4 The "live high-train low" model of altitude training.

Reprinted, by permission, from B.D. Levine and J. Stray-Gundersen, 1997, "Living high-training low: Effect of moderate-altitude acclimatization with low-altitude training on performance," *Journal of Applied Physiology* 83: 103.

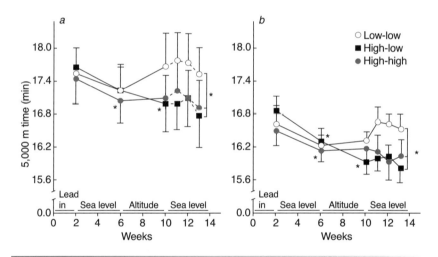

Figure 3.5 Time trial (5,000 m) results for all subjects *(a)* and for men only *(b)* at baseline, after sea level training in Dallas (sea level), and after either altitude training camp or sea level control (sea level). * Significantly different versus previous time point ($p < 0.05$); * next to brackets indicates a significant difference between groups ($p < 0.05$).

Reprinted, by permission, from B.D. Levine and J. Stray-Gundersen, 1997, "Living high-training low: Effect of moderate-altitude acclimatization with low-altitude training on performance," *Journal of Applied Physiology* 83: 110.

Following a four-week baseline period at sea level, the runners completed a 28-day training period in which they lived and completed their base training at 2,500 m/8,200 ft (Park City, UT) but did their interval workouts at 1,250 m/4,100 ft (Salt Lake City, UT). Serum EPO was evaluated within 20 hr after arrival at 2,500 m, and hemoglobin and hematocrit were measured within 20 hr postaltitude.

Compared with prealtitude values, significant increments were observed in serum EPO (92%), hemoglobin (8%), and hematocrit (4%). In addition, the runners produced an average 1% improvement ($p < 0.05$) in 3,000-m run time upon return to sea level, an improvement equivalent to 5.8 sec. The improvement in 3,000-m running performance was accompanied by a significant 3% improvement in maximal oxygen consumption. Stray-Gundersen et al. (2001) concluded that HiHiLo altitude training was as effective as LHTL altitude training in bringing about beneficial hematological changes and enhancing sea level endurance performance. In addition, this study demonstrated that HiHiLo altitude training led to improvements in sea level endurance performance among elite athletes, which was a unique finding in relation to previous studies evaluating well-trained but non-elite runners. It should be noted, however, that the conclusions of this study have come under criticism because it lacked a sea level control group.

As described earlier in this chapter, it appears that there is significant individual variation in the physiological responses of athletes upon exposure to altitude. Chapman et al. (1998) examined this topic within the context of LHTL altitude training. Following a four-week baseline period at sea level, female and male collegiate runners were divided into three training groups (n = 13 per group) and proceeded to complete a 28-day training period adhering to one of the following three training scenarios. An LHTL group lived at 2,500 m/8,200 ft (Park City, UT) and trained at 1,250 m/4,100 ft (Salt Lake City, UT). An LHTH group lived and trained at 2,500 m (8,200 ft). A HiHiLo group lived and completed their base training at 2,500 m, but did their interval workouts at 1,250 m. Serum EPO was evaluated within 30 hr after arrival at 2,500 m, whereas RBC mass, hemoglobin, and hematocrit were measured within 48 hr postaltitude.

Using a retrospective analysis, Chapman et al. (1998) classified the subjects from all three training groups as "responders" or "nonresponders" based on their performance in a postaltitude 5,000-m run. On average, the responders demonstrated a significant 4% improvement (37 sec) in the postaltitude 5,000-m run versus their prealtitude performance, whereas the nonresponders were approximately 1% slower (14 sec). Hematological data showed that the responders had a significantly larger increase in serum EPO (52%) compared with the nonresponders, who demonstrated a 34% increase in serum EPO. Similarly, postaltitude RBC mass for the responders was 8% higher (p < 0.05), whereas the nonresponders' RBC mass was only 1% higher (NSD), compared with prealtitude values (figure 3.6).

In addition, the responders showed a significant improvement in sea level $\dot{V}O_2$max (6%), whereas the nonresponders showed no change. A breakdown of the responders indicated that 47%, 18%, and 35% of them

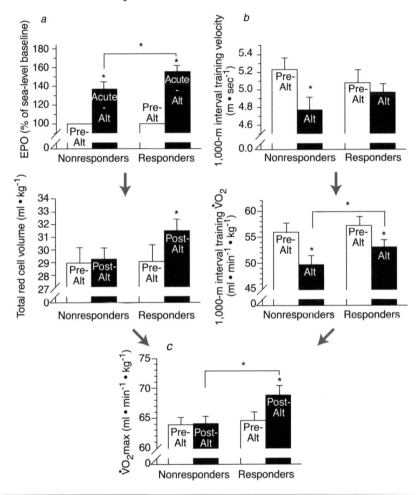

Figure 3.6 *(a)* Acclimatization responses (serum erythropoietin [EPO]; total red cell volume) of responders and nonresponders to altitude. *(b)* Training responses (1,000-m interval training velocity; 1,000-m interval training oxygen uptake [$\dot{V}O_2$]) of responders and nonresponders to altitude. *(c)* Changes in postaltitude maximal oxygen consumption ($\dot{V}O_2$max) of responders and nonresponders to altitude.
* Significant difference versus prealtitude (p < 0.05); * next to brackets indicates a significant difference between groups (p < 0.05).

Reprinted, by permission, from R.F. Chapman et al., 1998, "Individual variation in response to altitude training," *Journal of Applied Physiology* 85: 1452.

came from the LHTL, LHTH, and HiHiLo training groups, respectively. Of note was the fact that most of the responders completed a training program including some degree of LHTL altitude training. On the basis of the fact that the responders came from all three of the training groups, Chapman et al. (1998) concluded that each athlete may need to adhere to an individualized altitude training program that allows for optimal physical training and hematological response. In other words, although it appears that most athletes obtain the best physiological and performance results by following an LHTL altitude training program (either LHTL or HiHiLo), some athletes may get similar results by using traditional LHTH altitude training, or no altitude training program at all. The authors recommended that athletes be evaluated via controlled laboratory measures (hypobaric chamber) to determine the optimal altitude at which to live and train and consequently achieve the most beneficial physiological and performance results.

In summary, the LHTL method of altitude training has been in use since the mid-1990s and serves as an alternative to traditional LHTH altitude training. Data from several studies suggest that LHTL altitude training may produce beneficial changes in serum EPO, RBC mass, and hemoglobin, which in turn may lead to improvements in postaltitude $\dot{V}O_2$max and endurance performance. A number of different training strategies can be used in conjunction with LHTL altitude training, including normobaric hypoxia via nitrogen dilution (nitrogen apartment), supplemental oxygen, and hypoxic sleeping units. These strategies are described in detail in chapter 6 ("Current Practices and Trends in Altitude Training"), as is a summary of the scientific studies that have evaluated their efficacy in enhancing hematological variables and sea level performance. Table 3.5 summarizes the key findings of the scientific literature pertinent to LHTL altitude training.

Summary

"Does altitude training improve sea level performance?" That question has been debated by athletes and coaches for several years and has been the focus of a large body of scientific research spanning over three decades, as outlined in this chapter. Incredibly, when all the objective evidence is weighed, we still cannot say unequivocally whether altitude training leads to improvements in sea level performance.

As described earlier in this chapter, there are a number of reasons as to why the scientific research has not been able to demonstrate conclusively the efficacy/non-efficacy of altitude training. One of the most important factors has been the lack of a sea level control group in many of the studies, thereby making their results debatable and inconclusive.

Table 3.5 Scientific Literature About the Use of LHTL Altitude Training for the Enhancement of Sea Level Performance

Author	Altitude (m/ft)	Subjects	SL control	Altitude exposure (days)	Δ EPO[a] (%)	Δ Reticulocytes[a] (%)	Δ RCM[b] (%)	Δ Hb[b] (%)	Δ $\dot{V}O_2max$[b] (%)	Postaltitude SL performance[b]
Dehnert et al. 2002	Live: 1,956/ 6,415 (13 hr · day⁻¹) Train: 800/ 2,625	Trained triathletes (n = 11)	Yes	14	30*	14 NSD	NR	1 NSD	7 NSD	NSD in maximal running velocity vs. prealtitude
Liu et al. 1998	Live: 1,980/ 6,495 (12 hr · day⁻¹) Train: SL	Female and male trained triathletes (n = 11)	Yes	14	NR	NR	NR	NR	NR	Left ventricular cardiac function significantly enhanced Increased SV and $\dot{Q}$
Hutler et al. 1998	Live: 2,000/ 6,560 Train: 800/ 2,625	Trained triathletes (n = 11)	Yes	14	↑*	NR	NR	−2 NSD	NR	NR
Levine and Stray-Gundersen 1997	Live: 2,500/ 8,200 Train: 1,250/ 4,100	Female and male collegiate runners (n = 13)	Yes	28	NR	NR	5*	9*	5*	1% improvement in 5-km TT = 13.4 sec* 5-km TT on SL day 7, 14, 21 NSD versus SL day 3 5-km TT improvement correlated to $\dot{V}O_2max$ increase (r = 0.65, p < 0.01)

Study	Altitude (m)	Subjects	Iron	Days	ΔEPO (%)					Performance
Stray-Gundersen et al. 2001	Live: 2,500/8,200 Interval train: 1,250/4,100 Base train: 2,500/8,200	Female and male elite runners (n = 22) U.S. NT	No	28	92*	NR	NR	8*	3*	3-km TT faster by 1% (5.8 sec)*
Chapman et al. 1998[c]	Live: 2,500/8,200 Train: 1,250/4,100	Female and male collegiate runners (n = 13)	NR	28	R = 52*	NR	R = 8*	R = 8*	R = 6*	R = 5-km TT faster by 4% (37 sec)*
	Live: 2,500/8,200 Interval train: 1,250/4,100 Base train: 2,500/8,200	Female and male collegiate runners (n = 13)	NR	28	NonR = 34*		NonR = 1 NSD	NonR = 9*	NonR = 0	NonR = 5-km TT slower by 1% (14 sec)*
	Live: 2,500/8,200 Train: 2,500/8,200	Female and male collegiate runners (n = 13)	NR	28						

Studies are listed by ascending "living" altitude.

Δ = change.
[a] Measured within 1 to 5 days after arriving at altitude.
[b] Measured within 1 to 7 days postaltitude.
[c] Refer to the text for an explanation of the study design and retrospective analysis.
* Significant difference versus prealtitude ($p < 0.05$).
# Statistical analysis not reported.

EPO = erythropoietin (mU · ml^{-1}); Hb = hemoglobin (g · dL^{-1}); LHTL = "live high-train low"; NonR = nonresponders; NR = not reported/measured; NSD = no statistically significant difference; NT = national team; Q̇ = cardiac output; R = responders; RCM = red cell mass (ml · kg^{-1}); SL = sea level; SV = stroke volume; TT = time trial; V̇O$_2$max = maximal oxygen consumption.

Another factor has been the variety of research designs resulting in different "dosages" of hypoxia (altitude, number of hours per day, number of days at altitude, etc.). Preexisting medical conditions such as iron deficiency, injury, infection, or overtraining may have also confounded the results of some altitude training studies.

At this point, a couple of obvious trends are emerging from the scientific literature on altitude training:

- There is a distinct shift away from traditional LHTH altitude training and toward LHTL altitude training.
- Recent well-designed, well-controlled studies using the LHTL model have suggested that LHTL altitude training can result in significant erythropoiesis, leading to improvements in sea level $\dot{V}O_2$max and endurance performance. This has been demonstrated in well-trained as well as elite endurance athletes.
- On the basis of initial studies, there appears to be great individual variability at altitude in terms of one's
  - erythropoietic response,
  - ventilatory response,
  - cardiopulmonary response,
  - immune response,
  - ability to do high intensity training, and
  - chronic adaptation and acclimatization.
- There is interest in the research of non-traditional physiological variables (e.g., skeletal muscle buffering capacity) that may be influenced to a similar or greater degree at altitude than traditional markers (e.g., RBC and hemoglobin), and thus may effectively influence sea level performance.
- A number of researchers are investigating specific genetic factors associated with altitude training, particularly those that differentiate responders from non-responders.
- There is interest in research of the effect of altitude training on "middle distance" performance, for example, 4,000-m track pursuit cycling event, 400-m and 800-m runs, 200-m swim, etc.
- Research is being conducted on some of the novel simulated altitude training devices and applications: nitrogen apartment, altitude tent, intermittent hypoxic training, supplemental oxygen, etc. The scientific literature pertinent to these devices is reviewed in detail in chapter 6, "Current Practices and Trends in Altitude Training."

Although the current scientific literature is not conclusive regarding the efficacy of altitude training for the enhancement of sea level performance, it has been and will continue to be a very dynamic, exciting, and controversial area of applied sport science research. In the future we can expect altitude training research to employ everything from simple heart rate monitors to cutting-edge genetic technology to help answer the elusive question, "Does altitude training improve sea level performance?"

References

Adams, W.C., E.M. Bernauer, D.B. Dill, and J.B. Bomar. 1975. Effects of equivalent sea-level and altitude training on $\dot{V}O_2$max and running performance. *Journal of Applied Physiology* 39: 262-266.

Asano, K., S. Sub, A. Matsuzaka, K. Hirakoba, J. Nagai, and T. Kawaoka. 1986. The influences of simulated altitude training on work capacity and performance in middle and long distance runners. *Bulletin of Institute of Health and Sport Sciences* 9: 195-202.

Ashenden, M.J., C.J. Gore, G.P. Dobson, and A.G. Hahn. 1999. "Live high, train low" does not change the total haemoglobin mass of male endurance athletes sleeping at a simulated altitude of 3000 m for 23 nights. *European Journal of Applied Physiology* 80: 479-484.

Bailey, D.M., and B. Davies. 1997. Physiological implications of altitude training for endurance performance at sea level: A review. *British Journal of Sports Medicine* 31: 183-190.

Bailey, D.M., B. Davies, L. Romer, L. Castell, E. Newsholme, and G. Gandy. 1998. Implications of moderate altitude training for sea-level endurance in elite distance runners. *European Journal of Applied Physiology* 78: 360-368.

Berglund, B. 1992. High-altitude training. Aspects of hematological adaptation. *Sports Medicine* 14: 289-303.

Berglund, B., S.J. Fleck, J.T. Kearney, and L. Wide. 1992. Serum erythropoietin in athletes at moderate altitude. *Scandinavian Journal of Medicine and Science in Sports* 2: 21-25.

Boning, D. 1997. Altitude and hypoxic training—a short review. *International Journal of Sports Medicine* 18: 565-570.

Burtscher, M., W. Nachbauer, P. Baumgartl, and M. Philadelphy. 1996. Benefits of training at moderate altitude versus sea level training in amateur runners. *European Journal of Applied Physiology* 74: 558-563.

Buskirk, E.R., J. Kollias, R.F. Akers, E.K. Prokop, and E.P. Reategui. 1967. Maximal performance at altitude and on return from altitude in conditioned runners. *Journal of Applied Physiology* 23: 259-266.

Chapman, R.F., J. Stray-Gundersen, and B.D. Levine. 1998. Individual variation in response to altitude training. *Journal of Applied Physiology* 85: 1448-1456.

Chung, D.-S., J.-G. Lee, E.-H. Kim, C.-H. Lee, and S.-K. Lee. 1995. The effects of altitude training on blood cells, maximal oxygen uptake and swimming performance. *Korean Journal of Science* 7: 35-46.

Daniels, J., and N. Oldridge. 1970. The effects of alternate exposure to altitude and sea level in world-class middle-distance runners. *Medicine and Science in Sports* 2: 107-112.

Dehnert, C., M. Hutler, Y. Liu, E. Menold, C. Netzer, R. Schick, B. Kubanek, M. Lehmann, D. Boning, and J.M. Steinacker. 2002. Erythropoiesis and performance after two weeks of living high and training low in well trained athletes. *International Journal of Sports Medicine* 23: 561-566.

Dill, D.B., and W.C. Adams. 1971. Maximal oxygen uptake at sea level and at 3,090-m altitude in high school champion runners. *Journal of Applied Physiology* 30: 854-859.

Fandrey, J., and W.E. Jelkmann. 1991. Interleukin-1 and tumor necrosis factor-α inhibit erythropoietin production in vitro. *Annals of the New York Academy of Science* 628: 250-255.

Faulkner, J.A., J.T. Daniels, and B. Balke. 1967. Effects of training at moderate altitude on physical performance capacity. *Journal of Applied Physiology* 23: 85-89.

Faulkner, J.A., J. Kollias, C.B. Favour, E.R. Buskirk, and B. Balke. 1968. Maximum aerobic capacity and running performance at altitude. *Journal of Applied Physiology* 24: 685-691.

Friedmann, B., J. Jost, T. Rating, E. Weller, K.U. Eckardt, P. Bartsch, and H. Mairbaurl. 1999. Effects of iron supplementation on total body hemoglobin during endurance training at moderate altitude. *International Journal of Sports Medicine* 20: 78-85.

Fulco, C.S., P.D. Rock, and A. Cymerman. 1998. Maximal and submaximal exercise performance at altitude. *Aviation Space and Environmental Medicine* 69: 793-801.

Fulco, C.S., P.D. Rock, and A. Cymerman. 2000. Improving athletic performance: Is altitude residence or altitude training helpful? *Aviation Space and Environmental Medicine* 71: 162-171.

Gore, C.J., N.P. Craig, A.G. Hahn, A.J. Rice, P.C. Bourdon, S.R. Lawrence, C.B.V. Walsh, P.G. Barnes, R. Parisotto, D.T. Martin, and D.B. Pyne. 1998. Altitude training at 2690m does not increase total haemoglobin mass or sea level $\dot{V}O_2$max in world champion track cyclists. *Journal of Science and Medicine in Sport* 1: 156-170.

Hahn, A.G. 1991. The effect of altitude training on athletic performance at sea level—a review. *Excel* 7: 9-23.

Hahn, A.G., R.D. Telford, D.M. Tumilty, M.E. McBride, D.P. Campbell, J.C. Kovacic, R. Batschi, and P.A. Thompson. 1992. Effect of supplemental hypoxic training on physiological characteristics and ergometer performance of elite rowers. *Excel* 8: 127-138.

Hakkinen, K., K.L. Keskinen, M. Alen, P.V. Komi, and H. Kauhanen. 1989. Serum hormone concentrations during prolonged training in elite endurance-trained and strength-trained athletes. *European Journal of Applied Physiology* 59: 233-238.

Hutler, M., R. Schick, C. Dehnert, J.M. Steinacker, and D. Boning. 1998. "Sleep high—train low," effects of the hypoxia component. *International Journal of Sports Medicine* 19: S17.

Ingjer, F., and K. Myhre. 1992. Physiological effects of altitude training on elite male cross-country skiers. *Journal of Sport Sciences* 10: 37-47.

Jensen, K., T.S. Nielsen, A. Fiskestrand, J.O. Lund, N.J. Christensen, and N.H. Sechar. 1993. High-altitude training does not increase maximal oxygen uptake or work capacity at sea level in rowers. *Scandinavian Journal of Medicine and Science in Sports* 3: 256-262.

Klausen, T., T. Mohr, U. Ghisler, and O.J. Nielsen. 1991. Maximal oxygen uptake and erythropoietic responses after training at moderate altitude. *European Journal of Applied Physiology* 62: 376-379.

Kuno, S., I. Mitsuhara, K. Tanaka, Y. Itai, and K. Asano. 1994. Muscle energetics in short-term training during hypoxia in elite combination skiers. *European Journal of Applied Physiology* 69: 301-304.

Levine, B. 2002. Intermittent hypoxic training: Fact and fancy. *High Altitude Medicine and Biology* 3: 177-193.

Levine, B.D., and J. Stray-Gundersen. 1997. "Living high-training low": Effect of moderate-altitude acclimatization with low-altitude training on performance. *Journal of Applied Physiology* 83: 102-112.

Liu, Y., J.M. Steinacker, C. Dehnert, E. Menold, S. Bauer, W. Lormes, and M. Lehman. 1998. Effect of "living high—training low" on the cardiac functions at sea level. *International Journal of Sports Medicine* 19: 380-384.

Mizuno, M., C. Juel, T. Bro-Rasmussen, E. Mygind, B. Schibye, B. Rasmussin, and B. Saltin. 1990. Limb skeletal muscle adaptations in athletes after training at altitude. *Journal of Applied Physiology* 68: 496-502.

Nieman, D. 1997. Immune response to heavy exertion. *Journal of Applied Physiology* 82: 1385-1394.

Nieman, D.C., J.C. Ahle, D.A. Henson, B.J. Warren, J. Suttles, J.M. Davis, K.S. Buckley, S. Simandle, D.E. Butterworth, O.R. Fagoaga, and S.L. Nehlsen-Cannarella. 1995a. Indomethacin does not alter the natural killer cell response to 2.5 hours of running. *Journal of Applied Physiology* 79: 748-755.

Nieman, D.C., A.R. Miller, D.A. Henson, B.J. Warren, G. Gusewitch, R.L. Johnson, J.M. Davis, D.E. Butterworth, J.L. Herring, and S.L. Nehlsen-Cannarella. 1994. Effects of high- versus moderate-intensity exercise on circulating lymphocyte subpopulations and proliferative response. *International Journal of Sports Medicine* 15: 199-206.

Nieman, D.C., S. Simandle, D.A. Henson, B.J. Warren, J. Suttles, J.M. Davis, K.S. Buckley, J.C. Ahle, D.E. Butterworth, O.R. Fagoaga, and S.L. Nehlsen-Cannarella. 1995b. Lymphocyte proliferation response to 2.5 hours of running. *International Journal of Sports Medicine* 16: 406-410.

Peronnet, F., G. Thibault, and D.-L. Cousineau. 1991. A theoretical analysis of the effect of altitude on running performance. *Journal of Applied Physiology* 70: 399-404.

Rahkila, P., and H. Rusko. 1982. Effect of high altitude training on muscle enzyme activities and physical performance characteristics of cross-country skiers. In *Exercise and sport biology*, edited by P.V. Komi, pp. 143-151. Champaign, IL: Human Kinetics.

Rusko, H.K., H. Kirvesniemi, L. Paavolainen, P. Vahasoyrinki, and K.P. Kyro. 1996. Effect of altitude training on sea level aerobic and anaerobic power of elite athletes. *Medicine and Science in Sports and Exercise* 28 (Suppl. 5): S124.

Saltin, B., C.K. Kim, N. Terrados, H. Larsen, J. Svedenhag, and C.J. Rolf. 1995. Morphology, enzyme activities and buffer capacity in leg muscles of Kenyan and Scandinavian runners. *Scandinavian Journal of Medicine and Science in Sports* 5: 222-230.

Stray-Gundersen, J., C. Alexander, A. Hochstein, D. deLemos, and B.D. Levine. 1992. Failure of red cell volume to increase to altitude exposure in iron deficient runners. *Medicine and Science in Sports and Exercise* 24 (Suppl.): S90.

Stray-Gundersen, J., R.F. Chapman, and B.D. Levine. 2001. "Living high-training low" altitude training improves sea level performance in male and female elite runners. *Journal of Applied Physiology* 91: 1113-1120.

Svedenhag, J., K. Piehl-Aulin, C. Skog, and B. Saltin. 1997. Increased left ventricular mass after long-term altitude training in athletes. *Acta Physiologica Scandinavica* 161: 63-70.

Svedenhag, J., B. Saltin, C. Johansson, and L. Kaijser. 1991. Aerobic and anaerobic exercise capacities of elite middle-distance runners after two weeks of training at moderate altitude. *Scandinavian Journal of Medicine and Science in Sports* 1: 205-214.

Telford, R.D., K.S. Graham, J.R. Sutton, A.G. Hahn, D.A. Campbell, S.W. Creighton, R.B. Cunningham, P.G. Davis, C.J. Gore, J.A. Smith, and D.M. Tumilty. 1996. Medium altitude training and sea-level performance. *Medicine and Science in Sports and Exercise* 28 (Suppl. 5): S124.

Terrados, N., J. Melichna, C. Sylven, E. Jansson, and L. Kaijser. 1988. Effects of training at simulated altitude on performance and muscle metabolic capacity in competitive road cyclists. *European Journal of Applied Physiology* 57: 203-209.

Vallier, J.M., P. Chateau, and C.Y. Guezennec. 1996. Effects of physical training in a hypobaric chamber on the physical performance of competitive triathletes. *European Journal of Applied Physiology* 73: 471-478.

Vasankari, T.J., H. Rusko, U.M. Kujala, and I.T. Huhtaniemi. 1993. The effect of ski training at altitude and racing on pituitary, adrenal and testicular function in men. *European Journal of Applied Physiology* 66: 221-225.

Wilber, R.L. 2001. Current trends in altitude training. *Sports Medicine* 31: 249-265.

Wolski, L.A., D.C. McKenzie, and H.A. Wenger. 1996. Altitude training for improvements in sea level performance. *Sports Medicine* 22: 251-263.

Chapter 4
PERFORMANCE AT ALTITUDE FOLLOWING ACCLIMATIZATION

Do athletes adapt after living or training for several days at altitude? In an effort to answer that question, a number of scientific investigations have evaluated changes in endurance performance and $\dot{V}O_2max$ at altitude following several days or weeks of altitude acclimatization. One should note that four of those studies (Buskirk et al. 1967; Faulkner et al. 1967, 1968; Pugh 1967) were conducted prior to 1968 and were designed, in part, to provide athletes and coaches with practical information to be used in preparation for the 1968 Mexico City Olympics (2,300 m/7,544 ft). Few recent studies have measured the effect of altitude acclimatization on endurance performance and $\dot{V}O_2max$ at altitude. The reason may be that few international competitions are currently held at altitude locations because of the negative effect of hypoxia on endurance performance. Some exceptions have included the 1995 World Championships in road cycling held in Bogota, Colombia (2,640 m/8,660 ft), and the 2002 Winter Olympic Games held in the Salt Lake City area (1,250-2,003 m/4,100-6,750 ft). After a brief review of the scientific literature pertinent to performance at altitude following acclimatization, this chapter describes the altitude acclimatization program used by the U.S. Olympic Team in preparation for the Salt Lake City Games.

Early Research (1967-1975)

One of the initial studies on the effect of altitude acclimatization on athletic performance was conducted by Faulkner et al. (1967). The subjects were 15 male collegiate swimmers who lived and trained for 14 days at 2,300 m/7,544 ft (Alamosa, CO). Swim performances in 100-yd, 200-yd, and 500-yd time trials were significantly slower by 2%, 5%, and 6%, respectively, after one day at altitude. On the third day at altitude, swim performance remained 2% (100 yd) and 4% (200 yd, 500 yd) below sea level times, and these values remained unchanged when the swimmers were evaluated on the 14th day at 2,300 m. Faulkner et al. (1968) reported greater decrements in running performance for five male collegiate runners who lived and trained for 42 days at Alamosa, Colorado

(2,300 m/7,544 ft). On the seventh day at altitude, running performance (1-3 miles) was 4% to 10% slower compared with prealtitude sea level times (statistics not reported). Similar decrements in running performance were reported at the midpoint and end of the 42-day altitude training period. Compared with prealtitude sea level values, treadmill $\dot{V}O_2$max was 17% lower in the runners after one week at 2,300 m, and remained depressed by 14% and 12% after 16 and 31 days, respectively, of altitude acclimatization. Adams et al. (1975) reported minimal altitude adaptation in 12 trained male distance runners following 20 days of living and training at 2,300 m/7,544 ft (U.S. Air Force Academy, Colorado Springs, CO). Compared with prealtitude sea level values, 2-mile run time was 7% and 5% slower (statistics not reported) at the beginning and end, respectively, of the 20-day altitude training period. A similar pattern was seen in the runners' $\dot{V}O_2$max, which was 17% ($p < 0.05$) and 15% lower (no statistically significant difference [NSD]) at the beginning and end, respectively, of the 20-day altitude training camp versus prealtitude sea level $\dot{V}O_2$max.

Daniels and Oldridge (1970) evaluated the running performance of six male elite distance runners from the U.S. national team after two days (ALT 1), three to four weeks (ALT 3-4), and five weeks (ALT 5) of living and training at 2,300 m/7,544 ft (Alamosa, CO). Five days of living and training/competition at sea level preceded the performance tests at ALT 3-4 and ALT 5. Average 1-mile run time was 8% (4:26), 5% (4:18), and 4% (4:15) slower than the prealtitude sea level time (4:06) when measured at ALT 1, ALT 3-4, and ALT 5, respectively (statistics not reported) (figure 4.1a). Similarly, average 3-mile run time was 10% (15:25), 9% (15:19), and 8% (15:11) slower than the prealtitude sea level time (14:02) when measured at ALT 1, ALT 3-4, and ALT 5, respectively (figure 4.1b). The runners' maximal oxygen consumption was also assessed during their training period at altitude. Average $\dot{V}O_2$max was 14% (63.6 ml $\cdot$ kg^{-1} $\cdot$ min^{-1}), 10% (67.0 ml $\cdot$ kg^{-1} $\cdot$ min^{-1}), and 11% (66.0 ml $\cdot$ kg^{-1} $\cdot$ min^{-1}) lower than the prealtitude sea level $\dot{V}O_2$max value (74.8 ml $\cdot$ kg^{-1} $\cdot$ min^{-1}) when measured at ALT 1, ALT 3-4, and ALT 5, respectively (figure 4.1c). Similar results were reported by Pugh (1967) in British elite distance runners over a four-week period of time at 2,300 m/7,544 ft (Mexico City).

A few studies have measured the effect of altitude acclimatization on running performance at relatively high elevations (>3,050 m/10,000 ft). Dill and Adams (1971) evaluated six champion high school male cross country runners who lived and trained for 17 days at 3,090 m/10,135 ft (White Mountain Research Station, CA). Compared with the prealtitude sea level value, treadmill run time to exhaustion decreased by 27% at altitude based on the average of six tests conducted on days 2, 4, 7, 10, 13, and 16. Treadmill $\dot{V}O_2$max was reduced by 18% (statistics not

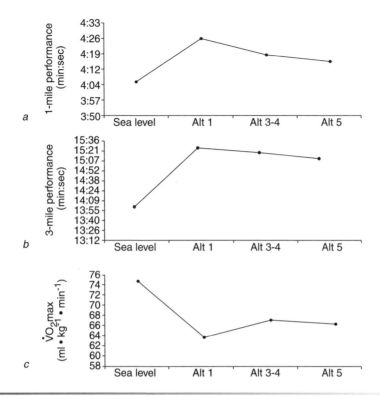

Figure 4.1 Changes in *(a)* 1-mile running performance, *(b)* 3-mile running performance, and *(c)* maximal oxygen consumption ($\dot{V}O_2$max) at altitude in elite U.S. male distance runners after altitude acclimatization at 2,300 m (7,544 ft). Alt 1 = after 2 days of altitude training; Alt 3-4 = after 3 to 4 weeks of altitude training; Alt 5 = after 5 weeks of altitude training.

Adapted, by permission, from J. Daniels and N. Oldridge, 1970, "The effects of alternate exposure to altitude and sea level in world-class middle-distance runners," *Medicine and Science in Sports* 2: 111.

reported) versus prealtitude sea level $\dot{V}O_2$max on the second day at 3,090 m. Maximal oxygen consumption remained depressed compared with prealtitude sea level values by 15% and 16% following 10 and 16 days of acclimatization, respectively. Figure 4.2 shows the time course and individual variability in $\dot{V}O_2$max in well-trained male distance runners over a 16-day altitude training period as reported by Dill and Adams (1971). In one of the initial studies of altitude acclimatization, Buskirk et al. (1967) evaluated six male collegiate runners who lived and trained for 63 days at an altitude of 4,000 m/13,120 ft (Nunoa, Peru). Average running times for the 440-yd, 880-yd, 1-mile, and 2-mile runs were 9%, 18%, 23%, and 19% slower versus prealtitude sea level times when measured after 40 to 57 days at altitude (statistics not reported). The

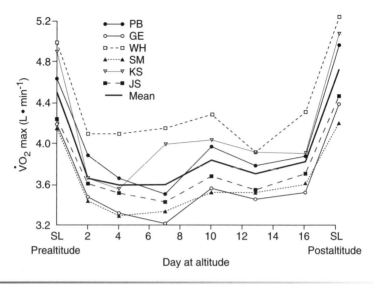

Figure 4.2 Individual variation in the changes in maximal oxygen consumption ($\dot{V}O_2$max) at altitude in six adolescent male distance runners after altitude acclimatization at 3,090 m (10,135 ft).

Reprinted, by permission, from D.B. Dill and W.C. Adams, 1971, "Maximal oxygen uptake at sea level and at 3,090-m altitude in high school champion runners," *Journal of Applied Physiology* 30: 856.

runners were also evaluated using a non-sport-specific cycle ergometer exercise protocol. Total ride time of the maximal cycle ergometer test was 12% less compared with prealtitude sea level values when measured on the third day at altitude. After 20 days of acclimatization, however, cycle ergometer ride time "approached prealtitude values." Compared with prealtitude sea level values, $\dot{V}O_2$max measured during the cycle ergometer test was 29% lower on the third day of the altitude training period. Maximal aerobic power remained 29% below sea level values on day 21, and improved only slightly to 26% of sea level $\dot{V}O_2$max after 48 days of acclimatization.

Recent Research (Post-1990)

As previously mentioned, there has been minimal research in the past decade on the effect of altitude acclimatization on athletic performance. This may be the case because few international-level competitions are held at altitude, and thus there appears to be less incentive to develop strategies for competing optimally at altitude. Nevertheless, a few recent studies have addressed the question. Both of the investigations described here involved elite-level athletes. Jensen et al. (1993) evaluated the effect of altitude acclimatization on the performance of nine male elite rowers from the Italian national team. The athletes lived and trained for 21 days at 1,822 m/5,975 ft (St. Moritz, Switzerland). During the initial days at altitude, work capacity (kilojoules) during a maximal rowing ergometer test and $\dot{V}O_2$max were below prealtitude sea level values by 7% ($p < 0.05$) and 15% ($p < 0.05$), respectively. At the end of the three-week altitude training camp, the rowers' work capacity at altitude improved slightly but was still 5% below ($p < 0.05$) prealtitude sea level values. A similar pattern was seen in the rowers' $\dot{V}O_2$max, which was higher following three weeks of altitude acclimatization relative to the initial days at altitude but was 10% below ($p < 0.05$) prealtitude sea level $\dot{V}O_2$max. A study by Vallier et al. (1996) examined the effect of intermittent hypoxic training on five female and male elite triathletes who were members of the French national team. The athletes trained three days per week for three weeks in a hypobaric chamber at a simulated altitude of 4,000 m (13,120 ft). Maximal power output measured during an incremental exhaustive cycle ergometer test was 28% lower ($p < 0.05$) than prealtitude sea level values upon acute exposure to simulated altitude, and remained unchanged over the duration of the 21-day training period. Cycle ergometer $\dot{V}O_2$max was 23% lower ($p < 0.05$) than at sea level upon acute exposure to a simulated altitude of 4,000 m. After 21 days of training in a hypobaric hypoxic environment, altitude $\dot{V}O_2$max improved only slightly, to a value that was 21% lower ($p < 0.05$) than prealtitude sea level $\dot{V}O_2$max.

Summary of the Scientific Research

Several scientific investigations have evaluated changes in endurance performance and $\dot{V}O_2$max at altitude following several days or weeks of altitude acclimatization. In general, the subjects recruited for the studies reviewed in this chapter were either well-trained or elite aerobic athletes who were evaluated on aerobic performance and $\dot{V}O_2$max within the initial days of exposure to actual or simulated elevations ranging from 1,822 m to 4,000 m (5,975-13,120 ft). In order to assess the effect of altitude acclimatization on aerobic performance, the athletes were also evaluated at select time points during altitude training periods that lasted from 14 to 63 days. Performance measures in these studies included sport-specific time trials, work capacity tests conducted on sport- and non-sport-specific ergometers, and endurance time during an exhaustive incremental exercise test.

Most of the investigations reviewed in this chapter showed decrements in aerobic performance (2-28%) and $\dot{V}O_2$max (14-29%) upon acute exposure to altitude. After several days or weeks of altitude acclimatization, aerobic performance (2-28%) and maximal oxygen consumption (8-26%) remained depressed compared with prealtitude sea level values. In many of the studies, however, there was a tendency toward improvement in aerobic performance and $\dot{V}O_2$max after several days of altitude acclimatization relative to the initial days at altitude. Collectively, these findings suggest that although aerobic performance and $\dot{V}O_2$max may remain depressed relative to sea level values regardless of the duration of acclimatization, they may improve after several days of altitude acclimatization compared with the initial days at altitude. Table 4.1 summarizes the results of several studies of the effect of altitude acclimatization on endurance performance and $\dot{V}O_2$max in trained aerobic athletes.

Additionally, the effect of altitude acclimatization on endurance performance at altitude (2,240-2,800 m/7,350-9,185 ft) is illustrated in figure 4.3 (Fulco et al. 1998, 2000). This figure includes data from four of the studies described in this chapter (Adams et al. 1975; Daniels and Oldridge 1970; Faulkner et al. 1967, 1968). As shown in the figure, for events lasting less than 2 min, there is minimal improvement in performance at altitude following 10 or more days of acclimatization. For events that are longer than 2 min in duration, however, there appears to be a minimum 2% improvement in endurance performance at altitude after chronic altitude acclimatization (>10 days). Collectively, these data suggest that altitude acclimatization (i.e., passive altitude exposure or altitude training) lasting 10 days or longer can lead to improvements in endurance performance at altitude in events lasting 2 to 16 min, and by extrapolation, in events lasting longer than 16 min (Fulco et al. 1998, 2000).

Practical Application

Daniels (1979) has estimated the approximate time loss for running distances at various altitudes for acclimatized runners. These estimations are presented in table 4.2 for running events lasting from 3 min to 4 hr at altitudes ranging from 1,000 m to 2,500 m (3,280-8,200 ft). Daniels (1979) strongly emphasized that athletes must take these altitude-induced time losses into account as they prepare their race strategy. For example, a 10,000-m race at 1,500 m above sea level (4,920 ft) may be up to 42 sec slower for international-level men. Thus, to be competitive the runner will have to race at a slower pace, particularly during the first half of the event. Modifications to race strategy may also be needed as to when the runner surges during the middle of the competition and when he or she kicks in the final lap of the race. The process of learning how to compete effectively at altitude via modifications to sea level race tactics is

Table 4.1 Scientific Literature About Changes in Aerobic Performance and $\dot{V}O_2$max at Altitude Following Acclimatization

Author	Altitude (m/ft)	Subjects	Altitude exposure (days)	Performance test	Acute performance decrement[a]	Decrement (%) in performance following acclimatization[b,c]	Acute $\dot{V}O_2$max decrement[a]	Decrement (%) in $\dot{V}O_2$max following acclimatization[b,c]
Jensen et al. 1993	1,822/ 5,975	Male elite rowers (n = 9) Italian NT	21	6-min rowing ergometer test	7*	5 (21)	15*	10* (21)
Faulkner et al. 1967	2,300/ 7,544	Male collegiate swimmers (n = 15)	14	100-yd TT	2*	2* (14)	NR	8* (14)
				200-yd TT	5*	4* (14)	NR	
				500-yd TT	6*	4 NSD (14)	NR	
Faulkner et al. 1968	2,300/ 7,544	Male collegiate runners (n = 5)	42	1-mile TT	4-10*	4-10*	17*	14* (12)
				2-mile TT				12* (31)
				3-mile TT				
Adams et al. 1975	2,300/ 7,544	Male trained runners (n = 12)	20	2-mile TT	7*	5* (20)	17*	15 NSD (20)
Daniels and Oldridge 1970[d]	2,300/ 7,544	Male elite runners (n = 6) U.S. NT	35	1-mile TT	8*	5* (21-28), 4* (35)	14*	10* (21-28), 11* (35)
				3-mile TT	10*	9* (21-28), 8* (35)		
Dill and Adams 1971	3,090/ 10,135	Male HS champion runners (n = 6)	17	TM run time to exhaustion	NR	27*[e]	18*	15* (10), 16* (16)

(continued)

Table 4.1 (continued)

Author	Altitude (m/ft)	Subjects	Altitude exposure (days)	Performance test	Acute performance decrement[a]	Decrement (%) in performance following acclimatization[b,c]	Acute $\dot{V}O_2$max decrement[a]	Decrement (%) in $\dot{V}O_2$max following acclimatization[b,c]
Buskirk et al. 1967	4,000/ 13,120	Male collegiate runners (n = 6)	63	440-yd TT	NR	9* (40-57)	29*	29* (21), 26* (48)
				880-yd TT	NR	18* (40-57)		
				1-mile TT	NR	23* (40-57)		
				2-mile TT	NR	19* (40-57)		
Vallier et al. 1996	4,000/ 13,120[f]	Female and male triathletes (n = 5) French NT	21	Cycling maximal power output	28*	28* (21)	23*	21* (21)

Studies are listed by ascending altitude.

[a] Acute performance and $\dot{V}O_2$max measured within 1 to 3 days after arrival at altitude. Exception: Faulkner et al. (1968) on day 7.
[b] All reported decrements in performance and $\dot{V}O_2$max are relative to prealtitude sea level values.
[c] Value in parentheses indicates the number of days at altitude at the time of the performance and $\dot{V}O_2$max tests.
[d] Refer to the text for explanation of the study design.
[e] Average of six tests conducted on days 2, 4, 7, 10, 13, and 16.
[f] Hypobaric chamber.
* Significant difference versus prealtitude sea level value ($p < 0.05$).
# Statistical analysis not reported.
HS = high school; NR = not reported/measured; NSD = no statistically significant difference; NT = national team; TM = treadmill; TT = time trial; $\dot{V}O_2$max = maximal oxygen consumption.

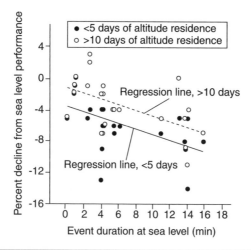

Figure 4.3 Effect of training at altitude on performance at altitude (2,240-2,800 m/7,350-9,185 ft). The x-axis is the event duration at sea level, and the y-axis is the percentage change from the sea level performance. The illustrated data are for the same individuals and events but for two different altitude durations—acute exposures (<5 days, filled circles) and chronic exposures (>10 days, unfilled circles). The regression lines depict the acute (solid line) and chronic (dashed line) exposure.

Reprinted, by permission, from C.S. Fulco et al., 2000, "Improving athletic performance: Is altitude residence or altitude training helpful?" *Aviation, Space, and Environmental Medicine* 71: 166.

called "competitive acclimatization" (Daniels 1979). However, competitive acclimatization can be acquired only after adequate physiological acclimatization has occurred.

Preparing for Competition at Altitude: The U.S. Olympic Team Experience at the 2002 Salt Lake City Winter Olympics

The United States enjoyed its greatest success in Olympic winter sport at the 2002 Salt Lake City Olympics. Team USA won an unprecedented 34 medals, finishing a close second to Germany in the total medal count. The United States' previous best had been 14 medals at the Lillehammer (1994) and Nagano (1998) Olympics. As U.S. athletes and coaches analyzed the success of Team USA at the Salt Lake City Olympics, one of the primary factors they cited was their thorough preparation for competing in the altitude environment of Salt Lake City (1,250 m/4,100 ft) and Park City (≥2,003 m/6,570 ft). For many athletes, this preparation started up to three years before the Salt Lake City Games. During that time, U.S. Olympic Committee sport physiologists Jay T. Kearney,

■ **Table 4.2 Potential Time Loss (min:sec) in Running Races Lasting 3 Minutes to 4 Hours for Acclimatized Runners**

Duration of sea level race	Altitude of 1,000 m (3,280 ft)	Altitude of 1,500 m (4,920 ft)	Altitude of 2,000 m (6,560 ft)	Altitude of 2,250 m (7,380 ft)	Altitude of 2,500 m (8,200 ft)
3 min	0:00.2	0:01.0	0:01.4	0:02.5	0:04.0
4 min	0:00.5	0:01.8	0:02.4	0:04.1	0:05.8
5 min	0:00.7	0:02.4	0:03.6	0:07.2	0:09.0
6 min	0:01.0	0:03.3	0:05.0	0:09.0	0:12.2
8 min	0:01.8	0:05.5	0:09.6	0:15.4	0:20.1
10 min	0:02.7	0:07.9	0:14.4	0:21.8	0:28.1
12 min	0:04.0	0:10.8	0:19.4	0:28.2	0:36.1
14 min	0:05.5	0:13.8	0:24.6	0:34.6	0:44.1
16 min	0:07.0	0:16.8	0:29.8	0:41.0	0:52.1
18 min	0:08.5	0:19.8	0:35.1	0:47.5	1:00.2
20 min	0:10.0	0:22.8	0:40.8	0:54.0	1:08.4

25 min	0:15.0	0:32.0	0:56.3	1:13.5	1:30.0
30 min	0:20.2	0:41.5	1:12.9	1:32.0	1:52.0
35 min	0:25.8	0:51.7	1:29.2	1:51.5	2:15.0
40 min	0:32.2	1:03.6	1:46.1	2:12.0	2:40.0
50 min	0:45.0	1:27.0	2:22.5	2:52.5	3:28.0
1 hr	1:00.0	1:52.0	3:00.0	3:36.0	4:17.0
1 1/2 hr	1:45.0	3:10.0	5:00.0	5:47.0	6:50.0
2 hr	2:30.0	4:30.0	7:00.0	8:00.0	9:30.0
2 1/2 hr	3:30.0	6:00.0	9:10.0	10:30.0	12:15.0
3 hr	4:30.0	7:30.0	11:20.0	13:00.0	15:00.0
3 1/2 hr	5:30.0	9:15.0	13:40.0	15:45.0	18:00.0
4 hr	6:30.0	11:00.0	16:00.0	18:30.0	21:00.0

Time values are expressed in minutes:seconds; unacclimatized runners might lose up to twice the expected time depending on pace and conditions.

Adapted, by permission, from J. Daniels, 1979, "Altitude and athletic training and performance," *American Journal of Sports Medicine* 7: 372.

Randy Wilber, Ken Rundell, and Mike Shannon worked closely with U.S. Olympic Team athletes and coaches from all sports for the purpose of designing and implementing the best possible altitude preparation program. Additional expertise was provided by altitude training experts Ben Levine, Jim Stray-Gundersen, Rob Chapman, Carl Foster, and Andy Subudhi. Essentially, three key issues were addressed in preparing U.S. athletes to compete successfully at the 2002 Salt Lake City Olympics:

- Acclimatization
- Training at altitude
- Away from training

Acclimatization

Two key questions emerged from U.S. athletes and coaches regarding altitude acclimatization prior to the Salt Lake City Games:

- How long?
- Where?

Table 4.3 describes the duration, location, and specific goals of the U.S. Olympic Team altitude preparation plan for each of the winter sports. The duration of altitude acclimatization varied depending on the physiological and technical demands of the sport. For example, the long-track speedskaters "lived high" in Park City (≥2,003 m/6,570 ft) and "trained low" at the Utah Olympic Oval speedskating rink (1,250 m/4,100 ft) in Salt Lake City for three years prior to the Games. The long-track speedskaters also used altitude tents (described in chapter 6, "Current Practices and Trends in Altitude Training") to help them maintain acclimatization effects when they traveled to lower elevations for competition and other activities. In contrast, the curlers arrived in Salt Lake City approximately two weeks prior to the Games.

One of the goals for every sport was acclimatization at the competition altitude, which in turn dictated where they would live. Another common goal was to become very familiar with the competition venues, thereby gaining a "home field advantage." Athletes in sports such as bobsled, alpine ski, freestyle ski, ski jumping, and speedskating had the additional goal of evaluating snow and ice conditions and aerodynamic factors. These factors also affected their decisions regarding the location and duration of altitude acclimatization.

Training at Altitude

The U.S. Olympic Team altitude preparation program addressed three key questions regarding training modifications at altitude:

Table 4.3 Altitude Acclimatization Program Used by U.S. Athletes in Preparation for the 2002 Salt Lake City Winter Olympics

Sport	Duration	Location	Goals
Endurance • Cross-country ski • Biathlon • Nordic combined • Long-track speedskating (distance)	Months-years	≥Park City (≥2,003 m/6,570 ft)	1a. Acclimatization at competition altitude 1b. Acclimatization at higher than competition altitude (live high-train/compete low) 2. Venue familiarization 3. Avoidance of distractions of Salt Lake City
Non-Endurance and Technical Group 1 • Bobsled, luge, and skeleton • Freestyle ski and snowboard • Alpine ski • Ski jumping • Long-track speedskating (sprint)	Several weeks (≥1 month)	Park City (2,003 m/6,570 ft)	1. Acclimatization at competition altitude 2. Venue familiarization 3. Evaluation of snow and ice conditions 4. Evaluation of aerodynamic factors 5. Avoidance of distractions of Salt Lake City
Non-Endurance and Technical Group 2 • Figure skating • Hockey • Short-track speedskating • Curling	2-week minimum*	Salt Lake City (1,250 m/4,100 ft) Ogden (1,463 m/4,800 ft)	1. Acclimatization at competition altitude 2. Venue familiarization 3. Evaluation of ice conditions 4. Evaluation of aerodynamic factors

Venues for the 2002 Winter Olympics were in Salt Lake City (1,250 m/4,100 ft), Ogden (1,463 m/4,800 ft), and Park City (≥2,003 m/6,570 ft).

* Trial run during the year prior to the Salt Lake City Olympics in Salt Lake City (1,250 m/4,100 ft), Ogden (1,463 m/4,800 ft), or Colorado Springs, CO (1,860 m/6,100 ft).

- Training volume
- Training intensity and velocity during interval workouts
- Recovery during interval workouts

In general, sea level training volume was reduced by 15% to 20% in the initial two weeks at altitude. Given that most U.S. Olympic Team sports used long-term acclimatization programs (>2 weeks), training volume was gradually increased to approximately 100% of sea level training volume by the fifth or sixth week at altitude. Training intensity or velocity during interval training workouts was reduced by 5% to 7% in the first two weeks at altitude and gradually increased toward 100% of sea level training intensity over several weeks. Replicating sea level training intensity during interval workouts was possible because the athletes took additional recovery between the work intervals. In general, sea level recovery time was doubled during interval workouts in the initial two weeks at altitude. Recovery time during interval workouts at altitude was maintained at 125% to 200% of sea level recovery throughout the acclimatization period to ensure good quality of effort.

The U.S. Olympic Team altitude preparation program also took into account the fact that there is great individual variability among elite athletes' responses at altitude. Many athletes were able to replicate sea level training volume and training intensity within a few weeks of altitude acclimatization with minimal negative effects. On the other hand, some athletes were unable to reproduce sea level workouts after several weeks at altitude and therefore did not strictly follow the guidelines just described. An acclimatization program was eventually designed for each athlete based on individual responses and needs.

A final point regarding the U.S. Olympic Team altitude preparation program should be mentioned. As described in chapter 5 ("Altitude Training Programs of Successful Coaches and Athletes"), several U.S. Olympic Team sports including long-track speedskating, cross-country ski, and Nordic combined used supplemental oxygen in preparation for the Salt Lake City Games. Utilization of supplemental oxygen allowed these athletes to "train low" without sacrificing the beneficial acclimatization effects obtained over several weeks or months at altitude.

Away From Training

In addition to modifications to training volume and training intensity, U.S. Olympic Team athletes adhered to several "non-training" guidelines in preparation for the 2002 Salt Lake City Olympics. Most of these steps were either a continuation or a modification of what the athletes normally did at sea level. However, some of these non-training guidelines were specific to altitude. For example, special effort was made to ensure that iron levels were sufficient to allow erythropoiesis to occur in athletes who committed to long-term altitude acclimatization (e.g., long-track speedskating). The importance of adequate iron stores prior to and during altitude acclimatization is discussed in chapter 2 ("Physiological Responses and Limitations at Altitude"). Regular blood testing was conducted to evaluate the athletes' iron status. Other non-training issues that received considerable attention were potential dehydration due to increased respiratory and urinary water loss, adequate carbohydrate and protein replacement, increased oxidative stress, adequate recovery and sleep, and increased susceptibility to illness. These issues are described in detail in chapter 2. Recommendations for non-training altitude issues are summarized in table 7.1 in chapter 7, "Recommendations and Guidelines."

Pre-Acclimatization

Several studies have suggested that pre-acclimatization may help athletes perform more effectively at altitude (Beidleman et al. 1997; Benoit et al. 1992; Geiser et al. 2001; Richalet et al. 1992; Savourey et al. 1994, 1998). Essentially, the process of pre-acclimatization exposes an individual to simulated altitude via a hypobaric chamber or inhalation of a hypoxic gas. Pre-acclimatization periods have been conducted at rest (5 days, 8 hr per day, 4,500 m to 8,500 m/14,760 ft to 27,880 ft) (Savourey et al. 1998), and in conjunction with a moderate-intensity cycling bout (3 weeks, 6 days per week, 2 hr per day, 4,500 m/14,760 ft) (Benoit et al. 1992) or a high-intensity cycling bout (6 weeks, 5 days per week, 30 min per day, 3,850 m/12,630 ft) (Geiser et al. 2001). In general, pre-acclimatization studies showed that subjects were able to adapt more quickly and perform more effectively at high altitude following a pre-acclimatization program. In addition, the pre-acclimatization regimen of Geiser et al. (2001) resulted in significant increases in mitochondrial volume density and capillary length density, both of which in theory could contribute to enhanced oxidative energy metabolism and aerobic performance at altitude. However, it should be noted that these pre-acclimatization studies were conducted on either untrained subjects or mountaineers preparing for treks to extreme altitude (e.g., Mount Everest 8,852 m/29,035 ft) and that objective athletic performance measures were not done.

Whether pre-acclimatization proves to be an effective strategy for athletes preparing for competition at moderate altitude (e.g., Salt Lake City 1,250 m/4,100 ft; Mexico City 2,300 m/7,544 ft) remains to be evaluated. It appears to hold great promise, particularly for athletes whose financial status or schedule prevents a prolonged (10-14 days) natural acclimatization period. However, it may be difficult for some athletes to get access to or afford the technology required for pre-acclimatization (hypobaric chamber, hypoxic gas). A more practical but somewhat less effective method of pre-acclimatization for athletes may be to reduce the duration of the recovery intervals during high-intensity workouts at sea level in the weeks prior to competition at altitude (Daniels 1979).

References

Adams, W.C., E.M. Bernauer, D.B. Dill, and J.B. Bomar. 1975. Effects of equivalent sea-level and altitude training on $\dot{V}O_2$max and running performance. *Journal of Applied Physiology* 39: 262-266.

Beidleman, B.A., S.R. Muza, P.B. Rock, C.S. Fulco, T.P. Lyons, R.W. Hoyt, and A. Cymerman. 1997. Exercise responses after altitude acclimatization are retained during reintroduction to altitude. *Medicine and Science in Sports and Exercise* 29: 1588-1595.

Benoit, H., M. Germain, J.C. Barthelemy, C. Denis, J. Castells, D. Dormois, J.R. Lacour, and A. Geyssant. 1992. Pre-acclimatization to high altitude using exercise with normobaric hypoxic gas mixtures. *International Journal of Sports Medicine* 13 (Suppl. 1): S213-S216.

Buskirk, E.R., J. Kollias, R.F. Akers, E.K. Prokop, and E.P. Reategui. 1967. Maximal performance at altitude and on return from altitude in conditioned runners. *Journal of Applied Physiology* 23: 259-266.

Daniels, J. 1979. Altitude and athletic training and performance. *American Journal of Sports Medicine* 7: 371-373.

Daniels, J., and N. Oldridge. 1970. The effects of alternate exposure to altitude and sea level in world-class middle-distance runners. *Medicine and Science in Sports* 2: 107-112.

Dill, D.B., and W.C. Adams. 1971. Maximal oxygen uptake at sea level and at 3,090-m altitude in high school champion runners. *Journal of Applied Physiology* 30: 854-859.

Faulkner, J.A., J.T. Daniels, and B. Balke. 1967. Effects of training at moderate altitude on physical performance capacity. *Journal of Applied Physiology* 23: 85-89.

Faulkner, J.A., J. Kollias, C.B. Favour, E.R. Buskirk, and B. Balke. 1968. Maximum aerobic capacity and running performance at altitude. *Journal of Applied Physiology* 24: 685-691.

Fulco, C.S., P.D. Rock, and A. Cymerman. 1998. Maximal and submaximal exercise performance at altitude. *Aviation Space and Environmental Medicine* 69: 793-801.

Fulco, C.S., P.D. Rock, and A. Cymerman. 2000. Improving athletic performance: Is altitude residence or altitude training helpful? *Aviation Space and Environmental Medicine* 71: 162-171.

Geiser, J., M. Vogt, R. Billeter, C. Zuleger, F. Belforti, and H. Hoppeler. 2001. Training high-living low: Changes of aerobic performance and muscle structure with training at simulated altitude. *International Journal of Sports Medicine* 22: 579-585.

Jensen, K., T.S. Nielsen, A. Fiskestrand, J.O. Lund, N.J. Christensen, and N.H. Sechar. 1993. High-altitude training does not increase maximal oxygen uptake or work capacity at sea level in rowers. *Scandinavian Journal of Medicine and Science in Sports* 3: 256-262.

Pugh, L.G.C.E. 1967. Athletes at altitude. *Journal of Physiology* 192: 619-646.

Richalet, J.P., J. Bittel, J.-P. Herry, G. Savourey, J.-L. LeTrong, J.-F. Auvert, and C. Janin. 1992. Use of a hypobaric chamber for pre-acclimatization before climbing Mount Everest. *International Journal of Sports Medicine* 13 (Suppl. 1): S216-S220.

Savourey, G., N. Garcia, Y. Besnard, A.-M. Hanniquet, M.-O. Fine, and J. Bittel. 1994. Physiological changes induced by pre-adaptation to high altitude. *European Journal of Applied Physiology* 69: 221-227.

Savourey, G., N. Garcia, J.-P. Caravel, C. Gharib, N. Pouzeratte, S. Martin, and J. Bittell. 1998. Pre-adaptation, adaptation and de-adaptation to high altitude in humans: Hormonal and biochemical changes at sea level. *European Journal of Applied Physiology* 77: 37-43.

Vallier, J.M., P. Chateau, and C.Y. Guezennec. 1996. Effects of physical training in a hypobaric chamber on the physical performance of competitive triathletes. *European Journal of Applied Physiology* 73: 471-478.

Part

Practical Application of Altitude Training

Chapter 5
ALTITUDE TRAINING PROGRAMS OF SUCCESSFUL COACHES AND ATHLETES

Several coaches and athletes have provided anecdotal evidence in support of the use of altitude training for the enhancement of sea level endurance performance. On the basis of several years of successful utilization of altitude training, these coaches and athletes have attempted to determine

- the optimal altitude at which to train,
- the optimal duration (number of weeks) to train at altitude,
- the best training regimen at altitude,
- when to return to sea level prior to competition, and
- how long the "altitude training effect" lasts after return to sea level.

It is important to emphasize that these accounts are supported by *anecdotal evidence*, rather than controlled scientific data. In other words, the recommendations provided by coaches and athletes in this chapter are based on how well athletes have responded, recovered, and performed during and after altitude training, as opposed to measurements of erythropoietin, red blood cell mass, and $\dot{V}O_2$max.

The altitude training programs described in this chapter are organized into three groups: (1) "traditional" altitude training camps, (2) training programs used by permanent altitude residents, and (3) contemporary altitude training programs that utilize "live high-train low" (LHTL) techniques. An emerging body of scientific literature suggests that LHTL may be the best method of altitude training (as described in chapter 3, "Performance at Sea Level Following Altitude Training," and chapter 6, "Current Practices and Trends in Altitude Training"), although the results of those studies are not entirely conclusive with respect to sea level performance. Some of the altitude training programs described in this chapter have utilized LHTL principles, and it is likely that more coaches and athletes will do so in the future. However, for practical reasons, LHTL altitude training may not be a viable option for many athletes because they may not have the financial means to access the geographic locales, laboratory facilities, or equipment necessary to conduct effective LHTL altitude training. In addition, it appears that some coaches and athletes are not convinced that LHTL altitude training is more effective than traditional altitude training in enhancing sea level performance.

Altitude Training Camps

Altitude training camps have been used for several years by athletes to prepare for competition at both sea level and altitude. Athletes typically move from sea level to moderate altitude, where they live and train for approximately four to six weeks. Altitude training camps are attractive in that they allow athletes to do very focused training in a beautiful mountain environment. On the other hand, altitude training camps may require the athlete to make a significant commitment in terms of money, travel, time away from job and family, and so on. This section focuses on coaches and athletes who have been successful in using altitude training camps, primarily in the sport of distance running.

Dr. Joe Vigil

Cross Country and Track Coach
Adams State College
Alamosa, Colorado

Dr. Joe Vigil served as coach of the highly successful Adams State College (Alamosa, CO) distance running program for 28 years (figure 5.1). During that time his athletes won 18 team and 89 individual national championships in cross country and track and field at the National Association of Intercollegiate Athletics (NAIA) and National Collegiate Athletic Association (NCAA) Division II levels of competition. Vigil is currently coaching a group of elite postcollegiate runners who live and train in Mammoth Lake, California (~2,439 m/8,000 ft). Runners he has guided over the years include U.S. Olympians Pat Porter and Deena Drossin. Dr. Vigil is recognized throughout the world as one of the foremost authorities on altitude training.

In addition to the collegiate runners who represent Adams State College, several runners from other countries have used Alamosa (2,300 m/7,544 ft) as an altitude training site for years. These athletes include Finland's Lasse Viren (gold medalist, 5,000 m and 10,000 m, Munich 1972 and Montreal 1976) and Italy's Gelindo Bordin (gold medalist, marathon, Seoul 1988). More recently, marathon runners from the highly successful Japanese and South Korean national teams have trained in Alamosa in preparation for international competition.

Figure 5.1 Coach Joe Vigil, former head cross country and track coach at Adams State College in Alamosa, Colorado (2,300 m/7,544 ft). On the right, Adams State records an unprecedented perfect score in winning the 1992 NCAA Division II National Cross Country Meet.

Reprinted, by permission, from J. Vigil, 1995, *Road to the top* (Albuquerque, NM: Creative Designs), viii.

Vigil suggests that altitude training be conducted at elevations of 2,100 m to 2,400 m (6,890-7,870 ft). He believes that proper utilization of altitude training can provide two benefits: (1) enhancement of general physical conditioning for the upcoming season and (2) preparation for a national or international championship. Vigil's altitude training program is divided into four phases:

- Acclimatization (1 week)
- Primary training following acclimatization (2-4 weeks)
- Recovery and preparation for return to sea level (1 week)
- Return to sea level

Figure 5.2 provides an overview of Vigil's altitude training program, showing the modifications in training volume, training intensity, and general strength training at various points in a six-week altitude training program. Vigil contends that individuals and teams who have positive experiences at altitude do so because they have a coach who understands the interrelationship between training volume and training intensity at altitude.

Acclimatization

The acclimatization phase is conducted over a period of approximately one week. During this phase, the athlete is exposed to as much open air activity as possible in order to facilitate the acclimatization process.

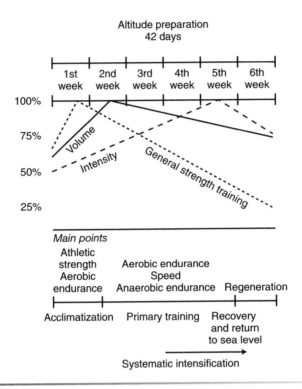

Figure 5.2 Selected factors of altitude training using a six-week period at altitude.

Reprinted, by permission, from J. Vigil, 1995, *Road to the top* (Albuquerque, NM: Creative Designs), 174.

Caution must be taken to ensure that the initial workouts are not too physically exhaustive. Vigil notes that the first week at altitude is the time when most athletes make the mistake of training too hard, thus preventing any positive adaptations from taking place in the subsequent weeks at altitude. Slow endurance runs of 30-min duration performed two to three times a day are recommended during the acclimatization phase. In addition, swimming, cycling, weight training, flexibility, and walking are included in the daily training schedule. Athletes should also make a concerted effort to hydrate as much as possible in order to avoid altitude-induced dehydration.

Primary Training Following Acclimatization

The primary training phase following acclimatization lasts between two and four weeks or longer, depending on the age and experience of the athlete. The first objective of the primary training phase is to increase the training volume toward 100% of sea level volume. This is generally accomplished during the first week of the primary training phase (figure

5.2). Elite male marathoners adhering to an eight-week altitude training program start off running 80 km (50 miles) during the first week, work up to running 225 km (140 miles) per week by the fourth week, and gradually taper back to 70 km (43.5 miles) by the final week. Olympic gold medalist Gelindo Bordin ran 257 to 354 km (160-220 miles) per week while training in Alamosa in preparation for the 1988 Seoul Olympic Games.

The second objective of the primary training phase is to modify and gradually increase the intensity of the training sessions. Vigil believes that the training intensity of aerobic workouts done at altitude must be slower by 0.3 to 0.4 m · sec^{-1} in comparison to similar aerobic workouts conducted at sea level. This difference in running velocity is equivalent to approximately 50 sec per 5,000 m, or 2 to 3 min per 10,000 m. For most athletes, anaerobic workouts at altitude must also be conducted at a reduced training intensity relative with similar anaerobic workouts done at sea level. Table 5.1 lists Vigil's recommended time reductions for interval training distances ranging from 200 m to 2,000 m. For example, if a workout requires the athlete to run 10 × 800 m, each work interval should be run 2 to 3 sec slower at altitude than the same workout at sea level. Vigil also suggests that athletes use longer recovery intervals during anaerobic workouts in order to ensure good quality of effort. The ultimate goal is to work as close as possible to sea level intensity during both the aerobic and anaerobic workouts conducted at altitude. The athlete moves progressively toward this goal by the final week of the primary training phase (figure 5.2). Vigil contends that a marathon runner training at altitude should be able to run 12 km to 15 km (7-9 miles) at sea level marathon pace, whereas a 10,000-m runner should be able to run 3 km to 5 km (2-3 miles) at a speed comparable to sea level 10,000-m pace. Many athletes coached by Vigil have achieved sea level training intensity in both aerobic and anaerobic workouts. These athletes, however, have typically completed two or three altitude training sessions per year for several years.

Table 5.1 Relationship Between Training Distance and Time Reduction at Altitude

Distance	Time reduction
200 m	0.5-1 sec faster
400 m	Same as at sea level
800 m	2-3 sec slower
1,000 m	4-6 sec slower
1 mile	7-9 sec slower
2,000 m	25-30 sec slower

Reprinted, by permission, from J. Vigil, 1995, *Road to the top* (Albuquerque, NM: Creative Designs), 175.

Recovery and Preparation for Return to Sea Level

The regeneration phase lasts for one week. The most important goal in this phase is for the athlete to avoid fatigue in the last four to five days before returning to sea level. During the regeneration phase there is a progressive decrease in training volume, intensity, and general strength training (figure 5.2). If the regeneration phase is carried out correctly, the transition back into sea level training should be smooth.

Return to Sea Level

The first four to five days after the return to sea level from an altitude training camp should be used for recovery and reestablishment of the normal sea level training pattern. Normal sea level training continues for the next two to three days. Vigil recommends that the first competition take place after six to eight days at sea level, although he notes that many outstanding sea level performances have been achieved as early as one to two days after the return from altitude. After the initial six to eight days at sea level, training consists of a significant reduction in training volume in combination with a systematic increase in training intensity in preparation for key national and international competitions. Table 5.2 provides a summary of Vigil's altitude training program.

Arturo Barrios

Cross Country and Track Coach
World Class Athlete Program
Boulder, Colorado

Arturo Barrios established himself as one of the world's best long distance runners during his career as an international competitor (figure 5.3). He represented Mexico in the 1988 Seoul Olympics and the 1992 Barcelona Olympics. In 1989, Barrios set a world record in the 10,000-m run with a time of 27:08.23. In addition, he was the 1989 International Amateur Athletics Federation (IAAF) world cross country champion at the 12-km distance. Barrios recently became a citizen of the United States. He currently lives in Boulder, Colorado (1,770 m/5,800 ft), where he coaches the World Class Athlete Program, a group of elite distance runners who are supported by the United States Army.

Barrios believes that altitude training can produce beneficial hematological changes such as an increase in red blood cell mass and hemoglobin concentration. These changes will lead to improvements in sea level competitive performance and will also allow athletes to train more effectively at sea level. In addition, Barrios believes that the hilly terrain of an altitude environment allows runners to develop the strength necessary for success in distance events, particularly the 5,000-m, 10,000-m, and marathon races. According to Barrios, runners in events from the 1,500

Table 5.2 Altitude Training Program of Joe Vigil

Phase	Duration	Training objectives
Acclimatization	1 week	• Slow endurance runs of 30-min duration are performed two to three times a day. • Swimming, cycling, weight training, flexibility, and walking are included in the daily training schedule.
Primary training following acclimatization	2-4 weeks	• Training volume is increased toward 100% of sea level volume. • For interval workouts, the pace of the work interval is slower and the duration of the rest interval longer compared with sea level pace and duration. • Training intensity for both aerobic and anaerobic workouts is gradually increased.
Recovery and preparation for return to sea level	1 week	• Fatigue should be avoided in the last 4 to 5 days before the return to sea level. • Training volume, training intensity, and general strength training are systematically decreased.
Return to sea level		• First 4 to 5 days are used for recovery and reestablishment of the normal sea level training pattern. • Normal sea level training continues for the next 2 to 3 days. • The first competition should take place after 6 to 8 days at sea level. • After the initial 6 to 8 days at sea level, training consists of a significant reduction in training volume in combination with a progressive increase in training intensity in preparation for key national and international competitions.

Used by elite distance runners in Alamosa, Colorado, at an elevation of 2,300 m (7,544 ft).

Figure 5.3 Arturo Barrios, International Amateur Athletics Federation world champion in cross country and former world-record holder in the 10,000-m run.

m to the marathon can benefit from altitude training, whereas athletes who compete in the 800 m or events of a shorter distance cannot benefit from altitude training because their speed training will be compromised at the higher elevations. Barrios' altitude training program is divided into three components:

- Acclimatization (1 week)
- Primary training following acclimatization (4-5 weeks)
- Return to sea level (2-10 days)

Acclimatization

The initial week at altitude is devoted to acclimatization. The athlete focuses on doing long runs of easy to moderate intensity. Sea level training volume is reduced by about 20% during the acclimatization period. For example, if the athlete was running 161 km (100 miles) a week at sea level, the training volume during the first week at altitude is reduced to 129 km (80 miles) a week. During the subsequent weeks at altitude, the training volume is gradually increased back to 100% of the sea level training volume. Interval training sessions are not attempted during the acclimatization phase.

Primary Training Following Acclimatization

The primary period of altitude training lasts four to five weeks. During this time, interval workouts are gradually added to the weekly training program. However, two important adjustments are made to the interval workouts. First, the pace at which the intervals are run at altitude is slower than the pace at which they were run at sea level. For example, if the athlete was running 5 × 1,600 m at 4:40 pace at sea level, then the 1,600-m intervals are run at 5:00 pace during the initial week of the primary altitude training period. Barrios recommends that the athlete run the 1,600-m intervals approximately 5 sec faster per week over the duration of the four- to five-week primary altitude training period. The second major modification to interval workouts is an increase in the amount of recovery between the work intervals. Barrios recommends that the rest interval be doubled during the initial week of the primary altitude training period. The duration of the rest interval is reduced gradually over the four- to five-week primary altitude training period, but the rest interval is never reduced to less than 3 min.

In addition to track interval sessions, hill workouts are introduced during the primary altitude training period. For example, a typical hill workout during the first week of the primary altitude training period might consist of 10 × 100-m repeats run on a grassy hill of moderate grade. A moderate-grade hill is recommended because it provides the runner with sufficient resistance for strength development, but is not so steep that it compromises the athlete's ability to maintain a smooth running cadence and rhythm. Over the four- to five-week primary altitude training period, the length of the hill repeats is increased gradually to approximately 400 m. Another example of a hill workout done in the Barrios program is a 10-mile distance run at an elevation of approximately 2,560 m to 2,590 m (8,400-8,500 ft) in the mountains outside Boulder. This 10-mile course consists of a series of rolling hills, and the workout is typically done once or twice a week at a moderate pace. The primary goal of this workout is strength development.

Return to Sea Level

According to Barrios, the optimal timing of return to sea level must be determined through trial and error and must be individualized for each athlete. What works for one athlete may be ineffective for another. Barrios recommends that each athlete try the following schedules to determine which one produces the best results: (1) return to sea level 1 to 2 days prior to competition, (2) return to sea level approximately 7 days prior to competition, and (3) return to sea level approximately 10 days prior to competition. It is important that the athlete experiment with each of these schedules well in advance of an important competition such as the World Championships or Olympics. At the height of his career, Barrios ran effectively in the 5,000-m and 10,000-m events regardless of whether he had returned to sea level 1 to 2 days or approximately 7 days prior to a major competition. In contrast, Barrios believes that

Table 5.3 Altitude Training Program of Arturo Barrios

Phase	Duration	Training objectives
Acclimatization	1 week	• Long distance runs of moderate intensity • Reduction of sea level volume by 20% • Interval training not attempted

Primary training following acclimatization	4-5 weeks	• Introduction of interval training:
		  • Pace of work interval must be slower than sea level pace.
		  • Duration of rest interval is doubled versus sea level rest interval.
		  • Gradual modifications are made to work interval pace (faster) and rest interval duration (shorter).
		• Weekly hill workouts:
		  • 10 × 100 m
		  • 10-mile run at 2,560-2,590 m (8,400-8,500 ft) over rolling terrain

Return to sea level	2-10 days	• Optimal schedule must be determined through trial and error:
		  • At 1-2 days prior to sea level competition
		  • At approximately 7 days prior to sea level competition
		  • At 7-10 days prior to sea level competition
		• Optimal schedule must be individualized for each athlete.
		• Performance may be enhanced for up to 2 weeks following return to sea level.

Used by elite distance runners in Boulder, Colorado, at an elevation of 1,770 m (5,800 ft).

marathon runners achieve their best results if they return to sea level 7 to 10 days prior to competition. On the basis of his own experience, Barrios contends that the beneficial effects of altitude training can last for approximately two weeks after return to sea level. Table 5.3 summarizes the altitude training program of Arturo Barrios.

Mark Plaatjes

Boulder, Colorado

Mark Plaatjes was born in South Africa but competed collegiately in the United States and became an American citizen in 1992. In 1993, he won the IAAF World Championship in the marathon (figure 5.4). Prior to his World Championship victory Plaatjes lived and trained in Boulder, Colorado, which is located at an elevation of 1,770 m (5,800 ft). He believes that altitude training is necessary in order to compete successfully at the international level in events ranging from the steeplechase to the marathon. Plaatjes believes that altitude training not only leads to improvements in sea level endurance performance, but also allows an athlete to establish a good aerobic base that in turn enhances the athlete's ability to train more effectively upon return to sea level. Plaatjes' altitude training program is divided into three components:

Figure 5.4 Mark Plaatjes, International Amateur Athletics Federation world champion in the marathon.

- Acclimatization (2-3 weeks)
- Primary training following acclimatization (4-6 weeks)
- Return to sea level (2-3 weeks)

Acclimatization

Similarly to others, Plaatjes believes that athletes should allow for a period of adjustment upon arrival at altitude. This acclimatization period lasts two to three weeks and consists of moderately intense aerobic work interspersed with a weekly hill workout. Plaatjes also recommends that athletes make a concerted effort to drink more water during the acclimatization period than they normally drink at sea level.

Primary Training Following Acclimatization

After the acclimatization period, altitude training lasts a minimum of four weeks, preferably six weeks. Plaatjes believes that the biggest mistake athletes make when they come to altitude is that they attempt to maintain or increase their sea level training volume. Thus, he recommends that athletes reduce their sea level training volume by 15% to 20%. For interval sessions, the intensity of the work interval is adjusted to a slower time, and athletes should double the amount of recovery time between work intervals in order to ensure good quality of effort.

In preparation for his gold medal performance in the marathon at the IAAF World Championships, Plaatjes ran an average of about 160 km (100 miles) per week and never exceeded 177 km (110 miles) per week. This is relatively low mileage for an international-caliber marathoner. Plaatjes did his aerobic training up in the mountains above Boulder at an elevation of 2,440 m to 3,050 m (8,000-10,000 ft) and completed his interval training sessions in the city of Boulder at 1,770 m (5,800 ft).

Return to Sea Level

Plaatjes believes that athletes who compete in the 5,000-m or 10,000-m track events need to return to sea level three weeks prior to peak competition. During the first week back at sea level, 5,000-m and 10,000-m runners should focus on speed work, which Plaatjes believes cannot be done effectively at altitude. Marathoners need only two weeks at sea level prior to competition because they do not need the high-intensity speed work required for success in the shorter and faster 5,000-m and 10,000-m track events. On the basis of his own experience and his observation of other world-class distance runners, Plaatjes believes that the beneficial effects of altitude training on sea level endurance performance can be maintained for up to two months following return to sea level. Table 5.4 provides a summary of Plaatjes' altitude training program.

Table 5.4 Altitude Training Program of Mark Plaatjes

Phase	Duration	Training objectives
Acclimatization	2-3 weeks	• Do not attempt to maintain or increase sea level training volume. • Perform moderately intense aerobic workouts. • Perform weekly hill workouts in preparation for more intense work.

Primary training following acclimatization	4-6 weeks	• Do not attempt to maintain or increase sea level training volume.
		• Reduce sea level training volume by 15-20%.
		• Anaerobic training:
		  • Gradually introduce anaerobic workouts.
		  • Workout intensity may have to be reduced.
		  • The duration of the rest interval is doubled.

Return to sea level	2-3 weeks	• 5,000-m and 10,000-m runners:
		  • Return to sea level 3 weeks prior to peak competition.
		  • Focus on speed work during the first week back at sea level.
		• Marathon runners:
		  • Return to sea level 2 weeks prior to peak competition.
		• Performance may be enhanced for up to 2 months following return to sea level.

Used in Boulder, Colorado, at an elevation of 1,770 m (5,800 ft).

Ørjan Madsen, PhD

Director, Olympiatoppen Altitude Project
Norwegian Olympic Committee

The Norwegian altitude training model is based on the experiences of Dr. Ørjan Madsen, who works as an advisor to the Norwegian Olympic Committee, the Norwegian Swimming Federation, and other Norwegian sport federations in the areas of altitude training, periodization, and physiological monitoring of elite athletes. His initial experience with altitude came when he competed as a swimmer for Norway in the 1968 Mexico City Olympics (elevation 2,300 m/7,544 ft). Madsen has extensive experience as both a sport scientist and coach and has worked with elite swimmers in Norway, Germany, and the U.S. Virgin Islands. He is well respected around the world for his ability to design and implement altitude training programs. Madsen was responsible for overseeing the operation of the "Altitude Project" in Norway, whose primary purpose was to prepare Norwegian winter sport athletes for the 2002 Salt Lake City Winter Olympic Games held at 1,250 m to 2,003 m (4,100-6,570 ft). The Norwegian team competed very successfully at the Salt Lake City Olympics, finishing third in the total medal count with 24 medals, including 11 gold. The Norwegians were particularly successful in the sport of cross-country skiing.

Initial Considerations in Designing an Altitude Training Program

Madsen believes that altitude training can enhance sea level performance when done for a sufficient period of time at the proper altitude. For elite athletes, altitude training should take place at specific points in the competitive season and works best when the training process is individually monitored. Madsen contends that in order to produce the desired physiological effects at altitude, training must adhere to the following guidelines (Madsen 1999):

1. The athlete must live and train at altitude for a minimum of two weeks.
2. The athlete must live and train at a minimum altitude of 1,800 m (5,905 ft).
3. Altitude training must be conducted for a minimum of 2 hr per day.
4. Coaches must account for individual differences among athletes such as
   - chronological and training age,
   - actual and absolute performance capacity,

- training phase (periodization), and
- previous altitude training experience.
5. Altitude training = "training at altitude."

The coach and athlete must consider all these factors carefully in preparation for an altitude training camp. One should never lose sight of the fact that altitude training is "training at altitude"; thus, scientific principles of training should be used in combination with a detailed periodization plan in the design of an altitude training program.

General Altitude Training Plan

Figure 5.5 shows the structure of a general altitude training plan used by Norwegian athletes. Specifically, it shows a 16-week training period leading up to the World Championships. The plan includes two altitude training camps, each of which is three weeks in duration. Working back from the World Championships, Altitude Camp 2 ends 15 to 21 days prior to the World Championships. The National Championships are typically held approximately eight weeks prior to the World Championships. This schedule allows athletes adequate time to recover from the National Championships and prepare for Altitude Camp 2. Altitude Camp 1 ends no later than seven to eight weeks prior to the National Championships, but can be scheduled earlier in the training year if necessary.

Specific Altitude Training Plan

Figure 5.6 provides a detailed look at the structure of a three-week altitude training camp. This 21-day altitude training period is divided into three phases:

- Adaptation (2 days)
- Application (17 days)
- Regeneration (2 days)

It is important that training in the first two days at altitude (adaptation phase), as well as the final two days at altitude (regeneration phase),

Altitude Camp 1	Sea level training	National Championship	Altitude Camp 2	Sea level training	World Championship
21 days	7-8 weeks	1-2 weeks prior to Altitude Camp 2	21 days	15-21 days	4-7 days

Figure 5.5 General altitude training program used by Norwegian athletes showing a 16-week training period including two altitude training camps.

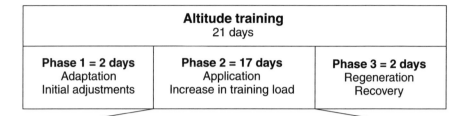

Figure 5.6 Structure of a three-week altitude training camp as used by Norwegian athletes.

be conducted at a low intensity. The initial two days are needed for acclimatization and initial adjustment, whereas the last two days are needed for recovery in preparation for return to sea level.

A total of 17 days (application phase) are devoted to actual training. The application phase can be divided into periods of basic training (7 days) and specific training (10 days). The primary objectives during the basic training period are general conditioning and aerobic training. Endurance 1 work is defined as steady-state training performed below the anaerobic threshold. Sprint training (<100 m) and cross/strength training should be according to individual needs.

In the specific training period, the training emphasis shifts to higher intensity. Endurance 1 training is maintained, and Endurance 2 training is added. Endurance 2 work is defined as training done at or above the anaerobic threshold. In addition, race pace workouts and other anaerobic training methods become important components in the specific training period. Sprint training and cross/strength training should be continued from the basic training period.

Return to Sea Level

Figure 5.7 provides a detailed look at the structure of sea level training following Altitude Camp 2, prior to the World Championships. Madsen believes that in order to achieve the desired results at the World Championships, the training carried out upon return to sea level is just as important as the training conducted at altitude. Madsen cautions that the optimal time frame for returning to sea level following altitude training can be determined for each athlete only through

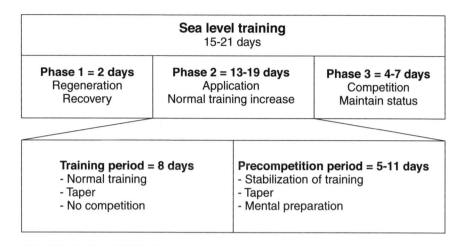

Figure 5.7 Structure of training used by Norwegian athletes after returning to sea level prior to a major competition.

trial and error. What works for one athlete may be totally ineffective for another.

The first two days of sea level training (regeneration phase) should be used for recovery; returning to sea level often involves extensive travel across several time zones. The third day at sea level often produces excellent results for single-day competitions, according to Madsen. However, the first eight days of the application phase (training period) are usually not well suited for a multiday competition. The athletes' physical response will be inconsistent during this eight-day period. Normal training is possible, but the results will vary. During the final 5 to 11 days of the application phase (precompetition period), the athlete enters a stabilization phase characterized by an increase in performance capacity. This period is ideal for competition, both single-day and multiday championships. According to Madsen, peak competitive performance can be expected from most athletes 16 to 24 days after the return to sea level.

Manfred Reiss, PhD

Head of Endurance Sport
Institute of Applied Training Science
Leipzig, Germany

Dr. Manfred Reiss serves as the head of the Endurance Sport Group at the highly respected Institute of Applied Training Science in Leipzig, Germany. In that capacity, he has worked for several years with Germany's elite endurance athletes on altitude training issues, focusing primarily on the use of altitude training camps in preparation for major international competition. Reiss has developed altitude training programs used by athletes preparing for competition at both sea level and altitude.

Altitude Training Camp in Preparation for Competition at Sea Level

Figure 5.8a shows the structure of a four- to five-week altitude training camp used to prepare for major competition at sea level. The structure comprises several phases (Reiss 2001):

- Preparation phase (4-6 days)
- Acclimatization phase (4-7 days)
- Load phase 1 (7-10 days)
- Regeneration phase 1 (2-3 days)
- Load phase 2 (7-10 days)
- Regeneration phase 2 (2-3 days)
- Reacclimatization phase (7-10 days)
- Postaltitude competition phase

The overall training load (volume + intensity) varies in each phase for the purpose of optimizing training.

Preparation Phase The preparation phase is conducted at sea level and lasts for four to six days. Athletes complete a physical exam during this time to identify any health factors that may preclude them from training at altitude. Training consists primarily of aerobic workouts. High-intensity training and competition, which require additional recovery and may therefore impair altitude acclimatization, are avoided.

Acclimatization Phase Most of the first week at altitude is devoted to gradual acclimatization. Workouts during this phase are essentially the same as during the preparation phase, that is, primarily aerobic training. Notice that the training load on the initial day at altitude is very light. At no time during the acclimatization phase does the training exceed a moderate level. Similar to the situation with the preparation phase, high-intensity interval training is not attempted during acclimatization.

Load Phase 1 and Regeneration Phase 1 During the first loading phase of the altitude training camp, the training load is gradually increased over the first few days to a relatively high level. The increase in training load is due primarily to an increase in volume versus intensity. Although the training load is classified as "high," minimal high-intensity interval training is attempted. The first loading phase is followed by two to three days of regeneration during which the training consists of low-volume aerobic workouts.

Load Phase 2 and Regeneration Phase 2 The second loading phase is similar to the first in duration (7-10 days). However, the training load in phase 2 increases by ≥50% compared with phase 1. This increase in training load is due primarily to an increase in training intensity, that is, a focus on very demanding "endurance" interval workouts. Clearly, this is the most important phase of the altitude training camp in terms of getting adequate nutrition and recovery in order to prevent illness, injury, or overtraining. Very close monitoring of the athlete is required during this time. The regeneration phase following load phase 2 is critical to allowing beneficial physiological training adaptations to occur.

Reacclimatization Phase Upon return to sea level, athletes complete a 7- to 10-day period of reacclimatization. Training during this time is relatively moderate and consists of both aerobic and anaerobic workouts. Anaerobic interval sessions are designed to restore race-pace leg speed ("turnover") that may have been compromised during the altitude training camp. These anaerobic interval sessions are typically done near the end of the reacclimatization phase with an emphasis on "quality" versus "quantity" of training.

Postaltitude Competition Phase The sea level competition phase begins approximately 7 to 10 days after the return from the altitude training camp. According to Reiss, the performance-enhancing effect is dependent on the duration of the altitude training camp and can last up to 30 days postaltitude. On the basis of his observations, Reiss contends that the time period following the second week back at sea level is very favorable for doing well in competition.

Altitude Training Camp in Preparation for Competition at Altitude

Contrasting to altitude training for the enhancement of sea level performance is the use of altitude training camps in preparation for major competitions held at altitude. A good example of this scenario was the 2002 Salt Lake City Winter Olympics. Several international teams conducted altitude training camps in the Salt Lake City/Park City area in preparation for the Games. Figure 5.8b shows the program that German athletes use to prepare for competition at altitude. Notice that the preparation and acclimatization phases are similar to those during the altitude training camp that prepares German athletes for sea level competition. However, the other phases are different as described in the following paragraphs.

Activation and Stabilization Phase The activation and stabilization phase follows the acclimatization phase and lasts from 7 to 10 days.

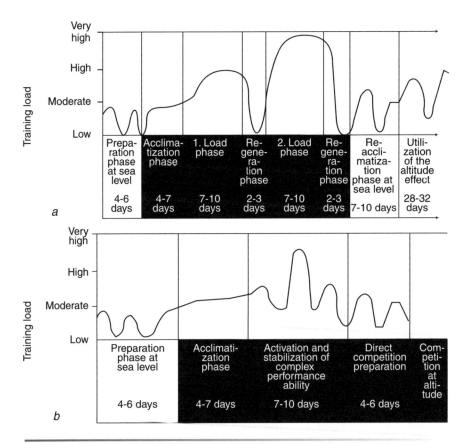

Figure 5.8 Phase structure of the German altitude training program for (a) training and performance improvement at sea level and (b) the preparation for competitions at altitude.

From Reiss 2001.

The primary goal is to activate and stabilize the athlete's sea level performance ability. This is accomplished via a progression of (1) aerobic endurance workouts ("moderate-high" training load), (2) endurance interval workouts ("very high" training load) similar to those in load phase 2 during the altitude training camp used in preparation for sea level performance, and (3) race pace efforts over short distances ("moderate-high" training load). One to two days of regeneration are placed between each of these three workouts.

Competition Preparation Phase Approximately four to six days are devoted to specific preparation for competition at altitude. The focus during this phase is on moderate aerobic training loads in combination with single-competition endurance loads and sport-specific neuromuscular development. In general, training volume and training intensity are low to moderate. There is a one- to two-day taper at the end of this phase as a final preparation for competition at altitude.

Training Programs Used by Permanent Altitude Residents

Some athletes such as the Kenyan distance runners were born and raised at altitude and continue to live at altitude. Other athletes live and train on a permanent or semipermanent basis in cities and at universities located at altitude. Some of these locations in the United States are the following:

- Colorado Springs, Colorado (U.S. Olympic Training Center and the U.S. Air Force Academy)
- Denver, Colorado (University of Denver)
- Boulder, Colorado (University of Colorado)
- Fort Collins, Colorado (Colorado State University)
- Alamosa, Colorado (Adams State College)
- Gunnison, Colorado (Western State College)
- Laramie, Wyoming (University of Wyoming)
- Flagstaff, Arizona (High Altitude Sports Training Complex and Northern Arizona University)
- Albuquerque, New Mexico (University of New Mexico)
- Salt Lake City, Utah (University of Utah)
- Provo, Utah (Brigham Young University)
- Park City, Utah

The altitude training programs described next have been used by athletes and coaches who are permanent altitude residents. These altitude training programs differ from traditional altitude training camps in that the athletes live and train at altitude most of the year, as opposed to going from sea level to altitude for limited periods of four to six weeks.

Jonty Skinner
National Team Director of Technical Support
USA Swimming
Colorado Springs, Colorado

Jonty Skinner currently serves as the National Team Director of Technical Support for USA Swimming and was formerly the head coach at the U.S. Olympic Training Center in Colorado Springs, Colorado. He is a former world-record holder in the 100-m freestyle event. Skinner has had great success as a coach as evidenced by the fact that his athletes have won 16 gold medals in four Olympic Games (1988-2000). Among those athletes is Amy Van Dyken (figure 5.9), who won four gold medals at the 1996 Atlanta Olympics, the best individual Olympic performance ever by a U.S. female athlete. Prior to the Atlanta Olympics, Van Dyken trained under Skinner in Colorado Springs at 1,860 m (6,100 ft). Other U.S. gold medalists coached by Skinner include B.J. Bedford, Ashley Tappin, Troy Dalbey, Angel Martino, and Jon Olsen.

Skinner's yearlong altitude training program is organized into three major phases:

- Altitude training period 1 (4 months)
- Altitude training period 2 (4 months)
- Sea level training period (4-6 weeks)

In Skinner's program, swimmers train for the initial eight months of the season at altitude in Colorado Springs, but periodically travel to sea level for competitions that last for only a few days. After this initial eight-month period at altitude, the swimmers move to sea level for the final four to six weeks of the season to work on speed

Figure 5.9 American Amy Van Dyken competing in the 1996 Atlanta Olympics, where she won four gold medals in swimming.

development and to prepare for the major national and international competitions.

Skinner believes that there are several advantages to this type of altitude training. These include an increase in red blood cell mass and hemoglobin concentration, as well as improved lactate tolerance. In addition, living and training at altitude will lead to enhanced recovery between rounds of competition at sea level and will allow for better recovery between repetitions, sets, and workouts after the athlete returns to sea level. Finally, Skinner believes that altitude training provides an important psychological advantage in that it forces athletes to deal with a higher level of physical pain during workouts.

Altitude Training Period 1

During the initial four months of the season, the swimmers live and train at 1,860 m (6,100 ft) in Colorado Springs. The primary goal of this training phase is to build a strong aerobic base for the upcoming season. Workouts focus on the development of general cardiovascular conditioning and strength. Lactate threshold workouts are introduced as the training phase progresses.

An interesting characteristic of Skinner's training program is that his athletes devote time during every workout to the development of alactate/anaerobic speed and power. This takes place in all of the three major training periods. The swimmer always does the alactate speed training at the beginning of practice and is typically required to do some type of supramaximal sprint lasting no longer than 8 sec. The sprint may be repeated using a 1:8 work-to-rest ratio. The swimmer performs the sprint in conjunction with a start, turn, or finish or while practicing the middle part of the race. Approximately 2% of the weekly training load is devoted to alactate speed and power training. Skinner believes that this type of limited but consistent emphasis on alactate training allows the athlete to maintain the neuromuscular aspects of speed and power without becoming overworked or overtrained.

Altitude Training Period 2

During the second four months of the season, the swimmers continue to live and train at altitude in Colorado Springs. With the swimmers having built a strong aerobic base in altitude training period 1, the training emphasis now shifts to the development of the swimmers' $\dot{V}O_2$max. As altitude training period 2 progresses, workouts that focus on the development of anaerobic endurance are gradually introduced. As in altitude training period 1, the swimmers devote a specific amount of training time at the beginning of every workout to the development of alactate speed and power.

Sea Level Training

During the final four to six weeks of the season, the swimmers in Skinner's program move down to sea level to do intense speed training. This type of training is not attempted at altitude because it may lead to overtraining. During the first week at sea level, the training emphasis is on recovery from the prolonged eight-month altitude training period, which allows the athlete to gradually adjust to the new sea level environmental conditions. Skinner recommends a minimum of one day of recovery per time zone traveled in the transition from altitude to sea level. Following this initial week of recovery and adaptation, the sea level swim workouts focus on the development of race-specific speed and technique. There are never more than two high-intensity workouts or competitions a week. During the final two to three weeks, the emphasis is on tapering and final preparation for national and international championships. Table 5.5 summarizes the altitude training program of Jonty Skinner.

Ronn Mann, Bobby McGee, and Mark Wetmore

Ronn Mann
Cross Country and Track Coach
Northern Arizona University
Flagstaff, Arizona

Bobby McGee
Cross Country, Track and Triathlon Coach
Boulder, Colorado

Mark Wetmore
Cross Country and Track Coach
University of Colorado
Boulder, Colorado

Ron Mann and Mark Wetmore are the head cross country and track distance coaches at Northern Arizona University (Flagstaff, AZ 2,165 m/7,100 ft) and the University of Colorado (Boulder, CO 1,770 m/5,800 ft), respectively. The collegiate and postcollegiate athletes whom they coach live and train at altitude on a permanent or semipermanent (≥9 months per year) basis. Both coaches oversee very successful NCAA Division I distance running programs for men and women. Between them they have coached five Olympians, 10 U.S. national champions, and over 200 NCAA Division I All-Americas. Their cross country teams have consistently placed in the top five at the NCAA Division I Championships, including first-place honors by the University of Colorado women (2000) and men (2001). Bobby McGee coaches national and international level distance runners and triathletes in Boulder, Colorado

Table 5.5 Altitude Training Program of Jonty Skinner

Phase	Duration	Training objectives
Altitude training 1	4 months	• Aerobic endurance and strength • Lactate threshold enhancement • Alactate speed and power development (2% of weekly training load)
Altitude training 2	4 months	• Maximal aerobic power ($\dot{V}O_2$max) • Anaerobic endurance • Alactate speed and power development (2% of weekly training load)
Sea level training	4-6 weeks	• During the first week, emphasis is on recovery from prolonged altitude training and adaptation to sea level environment. • After the first week, emphasis is on development of speed and refinement of swim technique. • Alactate speed and power development (2% of weekly training load). • The first competition should take place after 6-8 days at sea level. • During the final 2-3 weeks, training consists of a significant reduction in training volume and training intensity in preparation for key national and international competitions.

Used by elite swimmers at the U.S. Olympic Training Center at an elevation of 1,860 m (6,100 ft).

(1,770 m/5,800 ft). Among the current and past athletes that he has coached are Josiah Thugwane (Republic of South Africa), the 1996 Olympic gold medalist in the marathon; Barb Lindquist (USA), the number-one ranked female triathlete in the world in 2002 and 2003 (figure 5.10); and Colleen De Reuck, runner-up at the 1997 New York City Marathon and bronze medalist at the 2002 World Cross Country Championships.

Both Mann and Wetmore are convinced that living and training at altitude provides their athletes with distinct physiological and psychological advantages that allow them to compete very effectively at sea level. However, both coaches also warn that altitude training can be "a curse as well as a blessing." This is particularly true for athletes who do not have a good training plan at altitude and try to replicate sea level workouts, which ultimately results in illness, injury, or over-training.

Figure 5.10 Barb Lindquist, the number-one ranked female triathlete in the world in 2002 and 2003.

Mann and Wetmore agree that athletes cannot expect to replicate sea level training at altitude, particularly during interval workouts. Because their athletes cannot run as fast at altitude, these coaches are concerned about the potential loss of neuromuscular training effects, that is, the loss of race pace leg speed or "turnover." Mann and Wetmore both realize that this loss of leg speed is a negative aspect of year-round altitude training. The two coaches address this problem differently. Coach Wetmore uses "shorter than sea level" work intervals. For example, if the workout at sea level is 5 × 1,600 m at race pace, in Boulder the athlete will break the workout down into several 300-m and 400-m intervals done at sea level race pace with full recovery between each work interval. Coach Mann uses a similar strategy during interval training sessions at Flagstaff. In addition, he has the luxury of being able to drive 45 min from Flagstaff (2,165 m/7,100 ft) to Camp Verde, Arizona, at an elevation of approximately 910 m (2,985 ft), where his athletes are able to run much faster than at Flagstaff and thus maintain their leg speed and turnover. Prior to major competitions,

Mann's runners drive approximately 2 hr to Phoenix (518 m/1,700 ft), where they are able to replicate sea level workouts.

As to guidelines and recommendations, Mann and Wetmore agree that if altitude training is to be effective, the athlete's living environment needs to be as stress free as possible. They admit that this can be a major challenge, particularly among incoming freshman athletes living away from home for the first time in their lives. Mann and Wetmore continually emphasize to their athletes the importance of proper nutrition, hydration, and uninterrupted sleep. In addition, both programs have blood chemistry tests done on a regular basis to monitor iron status, potential anemia, and general health.

Neither coach has a definite answer to the question, "When is the best time to compete at sea level after coming down from altitude?" The primary reason is that the NCAA restricts travel to competitions to less than 48 hr prior to an event. Consequently, Mann and Wetmore have been unable to experiment with different timing strategies with their collegiate athletes. However, both coaches tend to believe that the optimal time to compete at sea level following altitude training is based on the duration of time spent at altitude. They also believe that the optimal timing of return to sea level is very individual and that it is best determined through systematic trial and error. Perhaps just as important to Mann and Wetmore is the issue of adapting to sea level heat and humidity. Because of their location in the mountains, Flagstaff and Boulder enjoy a relatively moderate-temperature, low-humidity climate most of the year. Mann and Wetmore believe that this is an important factor for their athletes to take into consideration as they prepare for sea level competition, and that 7 to 10 days should be allowed for acclimatization prior to competition in a hot and humid location. Again, because of NCAA travel restrictions, this precaution is not possible with the collegiate athletes, but is relevant to the postcollegiate runners these coaches work with.

Bobby McGee's approach to training athletes at altitude is similar to that of Ron Mann's and Mark Wetmore's. He uses shorter work intervals and longer recovery periods during high-intensity track workouts. Another workout that McGee uses effectively with his runners to help maintain leg speed and turnover is downhill running, typically done on a smooth, grassy surface with a grade of 5% to 7%. This type of downhill running forces the athlete to run relatively fast thereby training the "fast/speed" neuromuscular components of the running motion without expending excessive energy in the hypoxic altitude environment.

For marathon runners training at altitude, one of the workouts that McGee uses is an extended, gradual (2%-3%) downhill run of up to 24 km (15 miles). This workout is run at sea-level marathon race pace and is designed to help the athlete both physically and psychologically.

Upon going to sea level to complete, the athlete will have confidence in running the prescribed marathon race pace and will not worry that he or she lost leg speed and turnover while training at altitude.

The Kenyans

As described in the introduction, the 1968 Mexico City Olympics was the first of several Olympic Games in which the middle and long distance events in men's track and field were dominated by altitude-based countries in East Africa, in particular Kenya. Kenya participated in the Olympics for the first time in 1956, when it sent a small group of runners and boxers to the Melbourne Olympic Games. Kenya won its first Olympic medal in track and field in 1964, when Wilson Kiprugut won a bronze medal in the 800-m event at the Tokyo Olympic Games. Four years later at the 1968 Mexico City Olympics, the Kenyans emerged as the dominant world power in distance running when they won seven medals, led by the performance of Kipchoge "Kip" Keino, who won a gold medal in the 1,500-m run and a silver medal in the 5,000-m run. In total, the Kenyan distance runners won three gold, three silver, and one bronze medal at Mexico City in events ranging from the 800 m to the 10,000 m. Kenyan dominance of international distance running has continued to the present day. Table 5.6 lists Kenyan medalists in the Olympic Games from Tokyo 1964 through Sydney 2000. Despite the fact that Kenya did not participate in the 1976 and 1980 Games for political reasons, no other country has attained the success that the Kenyans have in Olympic middle and long distance running, although Ethiopia, another East African altitude country, has challenged the Kenyans over the years for world dominance in distance running, most recently at the 2003 IAAF World Championships in Paris.

One distance event that has been dominated by the Kenyan athletes is the 3,000-m steeplechase. Since the 1968 Mexico City Olympics, Kenyan runners have won the gold medal in this event in every Olympics in which they have participated. In addition, they have won both the gold and silver medals in the steeplechase in every Olympic Games they have participated in since 1968 with the exception of the 1984 Los Angeles Olympics. The Kenyans swept *all three medals* in the 3,000-m steeplechase at the 1992 Barcelona Olympics.

In addition to their great success in Olympic track and field, the Kenyans have dominated the IAAF World Cross Country Championships (figure 5.11). The Kenyan senior men have won the team championship for the last 16 years (1988-2003) and have never finished lower than fourth since they began participating in 1981. Since 1986, the Kenyan runners have failed to produce the individual champion in the senior men's division only four times. A similar record of success has been produced by the Kenyan junior men, as well as the Kenyan senior and junior women. Across all four divisions, the Kenyans have won a total

Table 5.6 Kenyan Medalists in Distance Running Events in the Olympic Games Since 1964

Olympics	Event	Medal	Kenyan athlete and result
1964 Tokyo	800 m	Bronze	Wilson Kiprugut 1:45.9
1968 Mexico City	800 m	Silver	Wilson Kiprugut 1:44.5
	1,500 m	Gold	Kip Keino 3:34.9
	3,000-m st	Gold Silver	Amos Biwott 8:51.0 Benjamin Kogo 8:51.6
	5,000 m	Silver Bronze	Kip Keino 14:05.2 Naftali Temu 14:06.4
	10,000 m	Gold	Naftali Temu 29:27.4
1972 Munich	800 m	Bronze	Mike Boit 1:46.0
	1,500 m	Silver	Kip Keino 3:36.8
	3,000-m st	Gold Silver	Kip Keino 8:23.6 Ben Jipcho 8:24.6
1976 Montreal*			
1980 Moscow*			
1984 Los Angeles	3,000-m st	Gold	Julius Korir 8:11.80
	10,000 m	Bronze	Michael Musyoki 28:06.46
1988 Seoul	800 m	Gold	Paul Ereng 1:43.45
	1,500 m	Gold	Peter Rono 3:35.96
	3,000-m st	Gold Silver	Julius Kariuki 8:05.51 Peter Koech 8:06.79
	5,000 m	Gold	John Ngugi 13:11.70
	10,000 m	Bronze	Kipkemboi Kimeli 27:25.16
	Marathon	Silver	Douglas Wakiihuri 2:10:47
1992 Barcelona	800 m	Gold Silver	William Tanui 1:43.66 Nixon Kiprotich 1:43.70
	3,000-m st	Gold Silver Bronze	Matthew Birir 8:08.84 Patrick Sang 8:09.55 William Mutwol 8:10.74
	5,000 m	Silver	Paul Bitok 13:12.71
	10,000 m	Silver	Richard Chelimo 27:47.72
1996 Atlanta	800 m	Bronze	Fred Onyancha 1:42.79
	1,500 m	Bronze	Stephen Kipkorir 3:36.72
	3,000-m st	Gold Silver	Joseph Keter 8:07.12 Moses Kiptanui 8:08.33
	5,000 m	Silver	Paul Bitok 13:08.16
	10,000 m	Silver	Paul Tergat 27:08.17
	Marathon	Bronze	Eric Wainaina 2:12:44

(continued)

Table 5.6 *(continued)*

Olympics	Event	Medal	Kenyan athlete and result
2000 Sydney	1,500 m	Gold	Noah Ngeny 3:32.07
		Bronze	Bernard Lagat 3:32.44
	3,000-m st	Gold	Reuben Kosgei 8:21.43
		Silver	Wilson Boit 8:21.77
	10,000 m	Silver	Paul Tergat 27:18.29
	Marathon	Silver	Eric Wainaina 2:10:31

* Kenya did not participate for political reasons. st = steeplechase.

Figure 5.11 Kenyan athletes competing in the International Amateur Athletics Federation World Cross Country Championships.

of 54 team championships (through 2003) with teams that have included 33 individual champions.

With very few exceptions, all of Kenya's distance runners have come from just four of the country's 40 tribes: Kikuyu, Kamba, Kisii (Gusii), and Kalenjin. About 75% of Kenya's best runners come from just one of these tribes, the Kalenjin, who compose only about 10% of the total Kenyan population. The Kalenjin live primarily on the western rim of the Rift Valley, which is the distinct geological formation running through Kenya in a north-south direction that separates the western third of Kenya from the rest of the country. The Kalenjin homeland is an area of rolling green hills located at an elevation ranging from 1,830 m to 2,440 m (6,000-8,000 ft).

Of course, the question that many people have asked since 1968 is "Why are the Kenyans so successful in distance running?" Several reasons have been proposed, including the following (Hamilton 2000; Noakes 2000):

- Living and training at altitude
- Genetic predisposition to success in endurance events
- Development of a high $\dot{V}O_2$max as a consequence of extensive walking and running at an early age
- Favorable muscle fiber composition and oxidative enzyme profile
- Unique viscoelastic property of the Achilles tendon and triceps surae muscle that promotes superior running biomechanics and physiological economy
- Psychological advantages

Some of these factors have been examined either scientifically or subjectively. The following sections provide some of the results of those studies and evaluations. In general, it appears that Kenyan distance running success is not based on some unique genetic or physiological characteristic but instead is a result of the optimal utilization of altitude in combination with moderate-volume, high-intensity training.

Physiological Factors Contributing to Kenyan Running Success

Scientific research has been conducted to determine if the Kenyan distance runners have a unique skeletal muscle composition or enzymatic profile that allows them to perform so well at middle and long distance events. Saltin et al. (1995a) performed a study in which they compared 13 Kenyan male elite distance runners with 12 Scandinavian male elite distance runners. Muscle fiber samples were obtained from the vastus lateralis and gastrocnemius muscles via needle biopsies. The results indicated that there were no statistically significant differences between the Kenyan and Scandinavian runners in terms of skeletal muscle fiber type and size, capillary density, or activity of the oxidative enzyme citrate synthase (CS). The only significant difference identified was in the activity of the fat-burning β-oxidative enzyme, β-hydroxyacyl-CoA-dehydrogenase (HADH), which was 20% higher in the Kenyan runners versus their Scandinavian counterparts. The authors speculated that the 20% higher HADH activity observed in the Kenyan runners was probably not due to their training, which was relatively high in intensity and thus did not favor lipid metabolism. The authors also reasoned that the difference in HADH activity was not due to the Kenyan runners' diet, which was relatively low in fat content. They concluded that the difference in HADH activity between the Kenyan and Scandinavian runners may have been due to genetic predisposition.

Increased HADH activity may offer a minor competitive advantage in the long distance events (10,000 m, half-marathon, marathon) because of the contribution of lipids to the total energy requirement in those races. However, lipid metabolism does not make a significant contribution to total energy demand in the middle distance events (800 m, 3,000-m steeplechase, 5,000 m). Therefore, given that the Kenyans have been successful over the full spectrum of middle and long distance running events, it does not seem logical to conclude that the 20% higher HADH activity identified in the Kenyans by Saltin et al. (1995a) can be considered a discriminating physiological factor that explains why the Kenyan runners have been so successful.

Many people have speculated that Kenyan distance runners have an exceptionally high $\dot{V}O_2$max that predisposes them to success in endurance events. The development of this relatively high $\dot{V}O_2$max has been attributed to the fact that most Kenyans are required to walk or run long distances in order to get to school or work. This relationship has never been demonstrated objectively. However, there is one scientific study whose purpose was to evaluate $\dot{V}O_2$max in Kenyan distance runners. Saltin et al. (1995b) assessed maximal oxygen consumption in six Kenyan male elite distance runners and six Scandinavian male elite distance runners at sea level and at 2,000 m (6,560 ft). Maximal oxygen consumption was similar in the Kenyan and Scandinavian runners regardless of whether they were tested at sea level (Kenyans = 79.9; Scandinavians = 79.2 ml · kg^{-1} · min^{-1}) or altitude (Kenyans = 66.3; Scandinavians = 67.3 ml · kg^{-1} · min^{-1}) (figure 5.12).

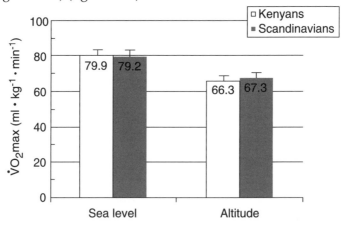

Figure 5.12 Maximal aerobic power ($\dot{V}O_2$max) at sea level and at altitude in Kenyan and Scandinavian elite male distance runners.

Reprinted, by permission, from B. Saltin et al., 1995, "Aerobic exercise capacity at sea level and altitude in Kenyan boys, junior and senior runners compared with Scandinavian runners," *Scandinavian Journal of Medicine and Science in Sports* 5: 213.

Saltin et al. (1995b) also reported a tendency (no statistical analysis) for the Kenyan runners to have better running economy (submaximal $\dot{V}O_2$ [ml · kg^{-1} · min^{-1}]) than the Scandinavians when running at race pace velocity at both sea level and altitude. The authors speculated that the enhanced running economy observed in the Kenyans may have been due in part to hill training, which in turn may have produced a better stride rate frequency in the Kenyan runners. The authors also suggested that the Kenyans' relatively good running biomechanics and physiological economy may have been affected by a unique viscoelastic component in the Achilles tendon and triceps surae muscle. This is a novel hypothesis that remains to be evaluated objectively. Another tendency that was observed (no statistical analysis) by Saltin et al. (1995b) was that blood lactate and blood ammonia concentrations were lower in the Kenyans than in the Scandinavians during running at race pace velocity at both sea level and altitude. The authors reasoned that since the Kenyans were residents of altitude, they experienced an attenuation in blood lactate accumulation, which in turn allowed them to maintain a higher relative exercise intensity before experiencing blood lactate accumulation.

Collectively, the results of the study by Saltin et al. (1995b) suggested that $\dot{V}O_2$max is not a factor that differentiates the performance of Kenyan distance runners from non-Kenyan distance runners. Although the Kenyan athletes had better running economy and lower blood lactate concentration than the Scandinavian athletes when running at race pace velocity, because of the lack of statistical analysis on these measures it is difficult to say whether they are viable discriminating factors. Additional research is warranted for the purpose of determining the relationship between running economy and blood lactate accumulation in Kenyan running performance.

In a recent study by Schmidt et al. (2002), total hemoglobin mass, red cell volume, and total blood volume were significantly higher in Columbian professional cyclists living and training at 2,600 m (8,530 ft) compared to German professional cyclists living and training at sea level. These results suggested that in endurance athletes who are native to moderate altitude, total hemoglobin mass and red cell volume are synergistically affected by living and training at altitude. The researchers who conducted this study speculated that, like the elite Columbian cyclists they evaluated, Kenyan distance runners may also have an enhanced RBC profile, and this may explain in part the great success of the Kenyans in endurance running. However, this theory is not conclusive because there have been no studies that have evaluated RBC parameters in the Kenyan runners specifically.

Training Factors Contributing to Kenyan Running Success

As previously noted, nearly all of Kenya's distance runners come from the Kalenjin tribe, which resides in the western third of the country at

Table 5.7 Altitude Training Locations of Kenyan Distance Runners

City/Town	Altitude	Kenyan athletes in training
Eldoret	2,100 m/6,890 ft	Moses Tanui 1991 world champion—10,000 m 1992 Olympic champion—800 m 1996 Boston Marathon champion
Eldama Ravine	2,100 m/6,890 ft	Lydia Cheromei 1991 junior world champion—cross country
Iten	2,300 m/7,545 ft	Benson Koech World junior record holder—800 m
Kamariny	2,400 m/7,870 ft	William Mutwol 1992 Olympic bronze medalist—3,000-m steeplechase
Kapsabet	1,950 m/6,395 ft	Ibrahim Hussein 1988, 1991, and 1992 Boston Marathon champion
Kaptagat	2,400 m/7,870 ft	FILA team
Machakos	1,600 m/5,250 ft	Cosmas Ndeti 1993, 1994, and 1995 Boston Marathon champion
Nandi Hills	2,000 m/6,560 ft	Peter Rono 1988 Olympic champion—1,500 m
N'gong Hills	2,000 m/6,560 ft	Paul Tergat 1995, 1996, and 1997 world champion—cross country 1996 Olympic silver medalist—10,000 m 2000 Olympic silver medalist—10,000 m
Nyahururu	2,350 m/7,710 ft	Moses Kiptanui 1991, 1993, and 1995 world champion—3,000-m steeplechase 1996 Olympic silver medalist—3,000-m steeplechase
Kipkabus	2,200 m/7,215 ft	Simon Chemoiywo

Adapted, by permission, from T. Tanser, 1997, *Train hard, win easy* (Mountain View, CA: Track and Field News Press), 77.

elevations ranging from 1,830 m to 2,440 m (6,000-8,000 ft). It is logical to assume that living and training at this altitude contributes in part to the success of the Kenyan runners. Although objective data to support this assumption are lacking, future scientific research may serve to elucidate the relationship between long-term altitude exposure and Kenyan running success. At present, the remarkable performance record of the Kenyan runners strongly suggests that living and training at moderate altitude is one factor that contributes to their athletic success. Most of Kenya's great runners have trained at several altitude training camps throughout western Kenya (table 5.7; Tanser 1997).

Although it stands to reason that long-term altitude exposure plays an important role in the success of the Kenyan distance runners, it also appears that training per se may be just as important. The Kenyans resent the implication that their success is based exclusively on some unique or inherent factor such as altitude, $\dot{V}O_2$max, skeletal muscle fiber characteristics, genetic predisposition, and so on. They are often frustrated that more attention is not paid to the fact that they train extremely hard and intelligently in preparation for international competition. Thus, the Kenyans claim that *hard training done at altitude* is the primary reason for their world domination in distance running. Kenyan coach Mike Kosgei puts it this way: "People think you have to be born a Kenyan to be a champion runner. If it were altitude alone then why are there not a lot of great runners from Nepal or Peru?" (Tanser 1997).

A closer look at Kenyan training patterns suggests that the distinguishing characteristic is not training volume, but training intensity. Table 5.8 (Saltin et al. 1995b) shows the weekly training schedule for Kenyan runners at St. Patrick's High School, which has produced several of Kenya's great male and female distance runners. The total volume run during the week is equivalent to 91 km to 139 km (56-86 miles), which is relatively similar to the total weekly volume of most European and U.S. runners of similar age, experience, and ability level. Of note, however, is the fact that the majority of the kilometers are run at a training intensity that is equivalent to 80% to 96+% of $\dot{V}O_2$max. A more recent study by Billat and colleagues (2003) provided additional evidence that Kenyan distance runners focus on moderate-volume, high-intensity training. These researchers reported that elite Kenyan men (n = 6) completed 4% of their total weekly training at a running speed that was faster than lactate threshold velocity (vΔ50) and completed 5% of their total weekly training at a running speed that was equivalent to $\dot{V}O_2$max velocity (v$\dot{V}O_2$max). For elite Kenyan women (n = 6), 4% of their total weekly training was done at vΔ50, whereas 8% was performed at v$\dot{V}O_2$max. It was also found that v$\dot{V}O_2$max was the best predictor of 10,000-m running performance for both the male and female Kenyan distance runners (Billat et al. 2003). This emphasis on moderate-volume, high-intensity training is reflected in quotes from some of Kenya's great runners (Tanser 1997):

Table 5.8 Weekly Training Schedule for Kenyan Runners at St. Patrick's High School

Day	Distance (km/miles)	Training intensity (% $\dot{V}O_2max$)			
		Low (~60)	Moderate (80)	High (90)	Interval (95+)
Monday A.M.	8-12/5-7.5	10	90	–	–
Monday P.M.	6-8/4-5	20	–	80	–
Tuesday A.M.	10-12/6-7.5	10	90	–	–
Tuesday P.M.	5-8/3-5	20	–	80	–
Wednesday A.M.	10-12/6-7.5	10	90	–	–
Wednesday P.M.	6-8/4-5	20	–	80	–
Thursday A.M.	8-12/5-7.5	10	90	–	–
Thursday P.M.	6-8/4-5	20	–	80	–
Friday A.M.	10-12/6-7.5	10	90	–	–
Friday P.M.	6-8/4-5	20	–	80	–
Saturday A.M.	Rest (or 9/5.5)	(40)	(60)	–	–
Saturday P.M.	6-12/4-7.5	10	–	–	90
Sunday A.M.	10-12/6-7.5	30	70	–	–
Sunday P.M.	Rest or 6/4	(10)	–	–	(90)

Reprinted, by permission, from B. Saltin et al., 1995, "Aerobic exercise capacity at sea level and altitude in Kenyan boys, junior and senior runners compared with Scandinavian runners," *Scandinavian Journal of Medicine and Science in Sports* 5: 217.

> *"Tempo training is practicing the pain we will face in competition; who wants to run slow in competition?"* Julius Ondieki
>
> *"If I feel good then I run faster no matter what the session. Don't waste good training time—if you feel good then run hard!"* John Ngugi
>
> *"There are no races in Europe as hard as the tempo runs we do here!"* Julius Korir

Thus, it appears that training intensity, not training volume, is an important distinguishing factor contributing to the Kenyans' success.

Summary

Kenyan athletes have dominated the sport of distance running since 1968, when they exploded onto the international scene at the Mexico City Olympics. Several reasons have been proposed to explain the success of the Kenyan distance runners, including altitude exposure, a high $\dot{V}O_2$max developed at an early age due to requisite walking and running, genetics, favorable skeletal muscle characteristics, and an enzymatic profile that favors endurance activity. A limited number of scientific studies have suggested that the Kenyan distance runners are similar to Western European runners in terms of skeletal muscle fiber type and number, oxidative enzyme profile, and $\dot{V}O_2$max. Preliminary data suggest that the Kenyans may be more biomechanically efficient and physiologically economical and may have a more favorable blood lactate response than their European counterparts; however, more research is needed in this area before definitive conclusions can be drawn.

It seems logical to assume that long-term altitude exposure contributes in part to the success of the Kenyan runners. However, it appears that the Kenyan training regimen, which emphasizes moderate-volume, high-intensity running, is also an important distinguishing factor contributing to the Kenyans' success. Thus, Kenyan distance running success is based primarily on living at altitude in combination with high-intensity training done at altitude. Other factors that probably contribute to the Kenyans' success include extensive postseason recovery periods (1-2 months), a climate that is favorable to outdoor training, and a relatively stable political and economic environment. In addition, the great tradition of Kenyan distance success serves to attract talented and motivated athletes to the talent pool. Distance running in Kenya is analogous to basketball in the United States or soccer in Brazil; it is a way for economically deprived athletes to improve the quality of their life. This situation breeds psychologically tough athletes. As Kenyan coach Mike Kosgei puts it, "Our success is based on *attitude*, not altitude."

Contemporary Altitude Programs That Utilize "Live High-Train Low" Techniques

"Live high-train low" (LHTL) altitude training is described in chapter 3 ("Performance at Sea Level Following Altitude Training"), as are several studies that support its use as a way to enhance sea level performance. Many contemporary athletes use LHTL in preparation for competition. Athletes can do this by living in LHTL locations such as Park City/Salt Lake City, Utah. In addition, athletes can do

LHTL altitude training by using simulated altitude devices such as nitrogen dilution, altitude tents, and supplemental oxygen. Given that most of these LHTL techniques involve either relatively expensive equipment or relocation to an LHTL-specific location, they are used primarily by elite-level athletes. This section presents examples of some LHTL altitude training programs used by elite athletes in preparation for Olympic competition.

USA—Preparation for the 2000 Sydney Summer Olympics

In preparation for the 2000 Sydney Olympics, U.S. Olympic Team athletes used LHTL altitude training at two of their national training centers—Chula Vista, California, and Colorado Springs, Colorado. The goal at each training center was to use LHTL to effectively prepare for Olympic competition held at sea level.

U.S. Olympic Training Center, Chula Vista, California

The U.S. Olympic Training Center in Chula Vista, California, is located at approximately 165 m (540 ft). In preparation for the Sydney Olympics, athletes training at the Chula Vista training center "slept high" in a simulated, normobaric hypoxic environment using altitude tents in their rooms. An altitude tent can be assembled in an athlete's bedroom. It creates an artificial altitude environment by reducing the amount of oxygen that enters the tent. Altitude tents are described in more detail in chapter 6, "Current Practices and Trends in Altitude Training." Athletes training at Chula Vista slept in the altitude tents for 8 to 10 hr a night at a simulated elevation of 2,800 m to 3,200 m (9,185-10,495 ft). Because the training center at Chula Vista is located at 165 m (540 ft), the athletes were able to "train low" very effectively with no travel or inconvenience.

Athletes followed this LHTL program for several months in advance of the Sydney Olympics. Some athletes slept in the altitude tents daily, whereas others used them in distinct blocks of two to three weeks. United States Olympic Team athletes from the sports of rowing, track and field, and canoe/kayak used this "sleep high-train low" strategy at the Chula Vista training center prior to the Sydney Olympics. An additional advantage of the Chula Vista location is that its relatively hot and sunny weather allowed U.S. Olympic Team athletes training there to gradually acclimatize to the environmental conditions they would encounter in Sydney.

U.S. Olympic Training Center, Colorado Springs, Colorado

The U.S. Olympic Training Center in Colorado Springs, Colorado, is located at approximately 1,860 m (6,100 ft), but has easy access to comfortable living conditions up to 2,745 m (9,000 ft). The LHTL strategy for U.S. Olympic Team athletes preparing for the Sydney Olympics in Colorado Springs was essentially the opposite of that used at the Chula Vista training site. Athletes in Colorado Springs "lived high" in a natural, hypobaric hypoxic environment at an elevation of 1,860 m to 2,745 m (6,100-9,000 ft). They were able to do their low- to moderate-intensity training at these altitudes. However, high-intensity interval training required the athletes to "train low" using supplemental oxygen. By breathing a medical grade gas composed of approximately 26% oxygen, athletes in Colorado Springs were able to artificially and temporarily go to "sea level" to complete their high-intensity workouts. This had the effect of allowing them to develop and maintain beneficial neuromuscular training effects (e.g., leg "turnover") that would be essential for them to compete optimally at sea level in Sydney. Supplemental oxygen training is described in more detail in chapter 6, "Current Practices and Trends in Altitude Training."

United States Olympic Team athletes from the sports of swimming, triathlon, cycling, and track and field used this LHTL strategy at the Colorado Springs training center prior to the Sydney Olympics. Because of the relatively moderate climate of Colorado Springs, the athletes spent 10 to 14 days at the Chula Vista training center immediately before traveling to Sydney to allow for adequate heat acclimatization.

Norway—Preparation for 2002 Salt Lake City Winter Olympics

The 2002 Winter Olympic Games were hosted by Salt Lake City, which is located at an altitude of 1,250 m (4,100 ft). Some of the events (biathlon, cross-country skiing, and the cross-country skiing portion of Nordic combined) were held in the Soldier Hollow area located in the mountains east of Salt Lake City at an elevation ranging from 1,670 m to 1,793 m (5,480-5,880 ft). In preparation for the Salt Lake City Winter Olympics, the Norwegians designed and implemented a detailed plan for their athletes. That plan is outlined in table 5.9. The three phases of the Norwegian altitude training plan for Salt Lake City were carried out in a continuous sequence. The overall goal was to prepare the Norwegian athletes for optimal performance at the Soldier Hollow cross-country ski venue.

During phase 1, the athletes lived and trained in Norway. There they used LHTL altitude training by living approximately 16 hr per day at a simulated altitude of 2,700 m (8,855 ft) in nitrogen-diluted apartments, and trained at 470 m (1,540 ft). The goals of phase 1 were to (1) acclimatize to altitude, (2) stimulate red blood cell production, and (3) train in a normobaric normoxic environment. During phase 2, the athletes moved to Utah, where they lived in the Deer Valley/Park

Table 5.9 Altitude Training Program Used by Norwegian Athletes for the 2002 Salt Lake City Winter Olympics

Phase	Duration	Training objectives
1	14 days • Live 16 hr · day^{-1} in "nitrogen house" at 2,700 m (8,855 ft) in Norway • Train at 470 m (1,540 ft) in Norway	• To acclimatize to altitude • To stimulate red blood cell mass • To train in a normal O_2 environment
2	14 days • Live at 2,500 m (8,200 ft) in Deer Valley, Utah • Train at 1,670 m to 1,793 m (5,480-5,880 ft) at the Soldier Hollow cross-country ski venue	• To adjust to the time change • To continue erythropoietic stimulation • To continue acclimatization • To train at the competition-specific altitude
3	16 days • Live at 1,700 m (5,575 ft) in Heber City, Utah • Train at 1,670 m to 1,793 m (5,480-5,880 ft) at the Soldier Hollow cross-country ski venue	• To be close to the competition venue • To live and train at the competition-specific altitude

City area (~2,500 m/8,200 ft) and trained at the Soldier Hollow cross-country ski venue (1,670-1,793 m/5,480-5,880 ft). This phase lasted two weeks; and the training objectives were to (1) adjust to the international time change, (2) continue erythropoietic stimulation, (3) continue the altitude acclimatization process, and (4) train at the competition-specific elevation of 1,670 m to 1,793 m (5,480-5,880 ft). The final phase lasted 16 days, during which time the athletes lived at approximately 1,700 m (5,575 ft) and trained at Soldier Hollow. This 16-day period preceded the 2002 Salt Lake City Olympics. The training goals of this final phase were to (1) live close to the competition venue and (2) live and train at the competition-specific elevation of 1,670 m to 1,793 m.

USA—Preparation for 2002 Salt Lake City Winter Olympics

The United States enjoyed its greatest success in Olympic winter sport at the 2002 Salt Lake City Olympics, winning 34 medals and finishing second in the total medal count. The program used to prepare U.S. athletes to compete optimally at the Salt Lake City Games is described in detail in chapter 4, "Performance at Altitude Following Acclimatization." Included in that altitude preparation program was the use of

LHTL altitude training by several sports; teams in all these sports were very successful at the Salt Lake City Olympics.

Three years before the Salt Lake City Games, the U.S. long-track speedskaters began living in Park City at approximately 2,440 m (8,000 ft) for the purpose of enhancing red blood cell mass and to acclimatize at an elevation significantly higher than the altitude of their competition venue in the Salt Lake City area (1,425 m/4,675 ft). The skaters performed dryland training of low to moderate intensity in Park City. They did high-intensity off-ice interval training using in-line skates on an oversized treadmill while using supplemental oxygen, which allowed them to artificially train low at "sea level." During the year prior to the Olympics, the long-track speedskaters had access to the Utah Olympic Park speedskating venue, thereby gaining valuable experience and knowledge of the ice conditions and aerodynamic factors. The athletes continued to live high in Park City and performed high-intensity on-ice workouts using a portable backpack supplemental oxygen unit (figure 5.13). Approximately three months prior to the Games, the long-track speedskaters refrained from using supplemental oxygen as they began their final preparation for the U.S. Olympic Trials and Olympic Games, both held at the Utah Olympic Park venue. The U.S. long-track speedskaters enjoyed unprecedented success in the Salt Lake City Olympics, with six athletes winning eight medals, including three gold medals, and achieving two world records. Their previous best had been eight medals (five gold) won by three athletes at the 1980 Lake Placid Olympics.

Several other U.S. teams used LHTL in preparation for the Salt Lake City Games. The U.S. Nordic combined skiers (ski jumping + cross-country ski) lived and trained year-round in Steamboat Springs, Colorado, at 2,096 m (6,875 ft) but frequently traveled to Park City to train at the Utah Olympic Park ski jump venue (2,098-2,235 m/6,880-7,330 ft) and the Soldier Hollow cross-country ski venue (1,670-1,793 m/5,480-5,880 ft). They used the portable supplemental oxygen system to train low during high-intensity on-snow workouts. During the year prior to the Olympics, the U.S. figure skaters used altitude tents and rooms to sleep high in their own homes while simultaneously training low at several ice rinks around the country, most of which were located at sea level. After the U.S. Olympic Trials, the figure skaters converged as a team in Colorado Springs (1,860 m/6,100 ft). They trained in Colorado Springs for the final two weeks prior to the Olympics before "dropping down" to compete at 1,250 m (4,100 ft) in Salt Lake City. The U.S. short-track speedskaters lived and trained year-round at the U.S. Olympic Training Center in Colorado Springs during the four years prior to the Salt Lake City Games. This allowed them to be fully acclimatized to the lower elevation of Salt Lake

Figure 5.13 Portable supplemental oxygen unit used by U.S. athletes in preparation for the Salt Lake City Winter Olympics.

City. The short-track speedskaters used regular domestic and international competitions as opportunities to train low. All three of these teams were very successful at the Salt Lake City Olympics, producing a total of six medals (two gold, two silver, two bronze) for Team USA.

Summary

This chapter describes the altitude training programs of several coaches and athletes who have been successful at the international level of competition in different sports. Many of these athletes and coaches have

Table 5.10 Guidelines and Recommendations for Altitude Training Camps

	Training guidelines and recommendations
Altitude	The optimal altitude at which to live and/or train appears to be between 1,800 m and 2,500 m (5,905-8,200 ft).
Duration	The optimal duration for an altitude training camp appears to be 3 to 6 weeks.
Training	Acclimatization (~1 week) Primary training following acclimatization (2-4 weeks): • Reduce sea level training volume by 10% to 20%. Gradually increase the training volume over the duration of the altitude training period by 3% to 5% per week. Some experienced athletes may be able to build up to their normal sea level training volume by the end of the altitude training period. • During interval training workouts, the pace of the work interval must be slower versus the sea level pace. The reduction in work interval pace should be determined for each athlete on an individual basis; however, it appears that a "safe" starting point would be a reduction of 5% to 7%. Gradually increase the pace of the work interval by 0.5% to 1% per week over the duration of the altitude training period. • During interval training workouts, the duration of the rest interval must be longer versus the sea level rest interval. The length of the rest interval should be determined for each athlete on an individual basis; however, it appears that a "safe" starting point would be to double the time of the rest interval. Gradually decrease the duration of the rest interval by 2% to 3% per week as the altitude training period progresses.
Return to sea level	The optimal timing of the return to sea level is variable. Therefore, an optimal schedule should be determined for each athlete through trial and error, experience, and performance results. What works for one athlete may be ineffective for another.
Duration of altitude training effect	It is difficult to estimate how long the benefits of altitude training last after an athlete returns to sea level. Some coaches and athletes believe that the beneficial effects of altitude training last approximately 2 weeks, whereas others are convinced that they last for up to 2 months following return to sea level.

Based on the anecdotal experiences of several coaches and athletes who have been successful at the international level of competition.

used "traditional" altitude training camps in preparation for major competition at sea level. General guidelines and recommendations from these coaches and athletes who have successfully used altitude training camps are summarized in table 5.10. In contrast to athletes who travel to altitude from sea level for periods of four to six weeks are those athletes who are permanent altitude residents. Examples of permanent altitude residents include the Kenyan distance runners and athletes living in altitude cities such as Colorado Springs, Colorado; Boulder, Colorado; Flagstaff, Arizona; and Park City, Utah. Finally, some athletes have the best of both worlds; that is, they "live high-train low" using devices such as nitrogen-diluted apartments, altitude tents, and supplemental oxygen. However, given the unique nature and cost of these LHTL devices, they are used primarily by elite, Olympic-caliber athletes.

References

Billat, V., P.-M. Lepretre, A.-M. Heugas, M.-H. Laurence, D. Salim, and J.P. Koralsztein. 2003. Training and bioenergetic characteristics in elite male and female Kenyan runners. *Medicine and Science in Sports and Exercise* 35: 297-304.

Hamilton, B. 2000. East African running dominance: what is behind it? *British Journal of Sports Medicine* 34: 391-394.

Madsen, Ø. 1999. Hypoxia—the "magic pill" to enhance performance in endurance sports in the 21st century. In *Proceedings of the Second Annual International Altitude Training Symposium.* Flagstaff, AZ.

Noakes, T.D. 2000. Physiological models to understand exercise fatigue and the adaptations that predict or enhance athletic performance. *Scandinavian Journal of Medicine and Science in Sports* 10: 123-145.

Reiss, M. 2001. Basic methodological principles of altitude training in elite sport. In *Proceedings of the Second Annual International Altitude Training Symposium.* Flagstaff, AZ.

Saltin, B., C.K. Kim, N. Terrados, H. Larsen, J. Svedenhag, and C.J. Rolf. 1995a. Morphology, enzyme activities and buffer capacity in leg muscles of Kenyan and Scandinavian runners. *Scandinavian Journal of Medicine and Science in Sports* 5: 222-230.

Saltin, B., H. Larsen, N. Terrados, J. Bangsbo, T. Bak, C.K. Kim, J. Svedenhag, and C.J. Rolf. 1995b. Aerobic exercise capacity at sea level and altitude in Kenyan boys, junior and senior runners compared with Scandinavian runners. *Scandinavian Journal of Medicine and Science in Sports* 5: 209-221.

Schmidt, W., K. Heinicke, J. Rojas, J.M. Gomez, M. Serrato, B. Wolfarth, A. Schmid, and J. Keul. 2002. Blood volume and hemoglobin mass in endurance athletes from moderate altitude. *Medicine and Science in Sports and Exercise* 34: 1934-1940.

Tanser, T. 1997. *Train hard, win easy.* Mountain View, CA: Tafnews Press.

Vigil, J. 1995. *Road to the top.* Albuquerque, NM: Creative Designs.

Chapter 6
CURRENT PRACTICES AND TRENDS IN ALTITUDE TRAINING

In recent years endurance athletes have started to use several novel strategies and devices for altitude training (Wilber 2001). These include

- normobaric hypoxia via nitrogen dilution,
- supplemental oxygen,
- hypoxic sleeping units, and
- intermittent hypoxic exposure.

This chapter provides an overview of these novel approaches to altitude training and reviews the scientific literature pertinent to each. A final section of the chapter briefly addresses the issue of the ethical integrity of using these devices.

Normobaric Hypoxia via Nitrogen Dilution

Normobaric hypoxia refers to simulated altitude that is created via a reduction in oxygen concentration, without any change in barometric pressure. "Nitrogen apartment" is a term used to describe a normobaric hypoxic apartment that simulates an altitude environment equivalent to approximately 2,000 m to 3,000 m (6,560-9,840 ft). The nitrogen apartment was developed by Finnish sport scientists in the early 1990s for the purpose of simulating an altitude environment. In Finland, the "nitrogen house" is in fact a series of well-furnished hotel rooms that provide a high level of comfort and privacy. In several countries, elite athletes have individual nitrogen apartments within their homes.

Athletes who use a nitrogen apartment adhere to "live high-train low" (LHTL) altitude training. Typically they live and sleep in a simulated altitude environment in the nitrogen apartment for 8 to 18 hr a day but complete their training at sea level, or approximate sea level conditions. The nitrogen apartment simulates elevations equivalent to approximately 2,000 m to 3,000 m (6,560-9,840 ft) via manipulation of the oxygen concentration within the apartment. The barometric pressure within the nitrogen apartment is equivalent to that at sea level

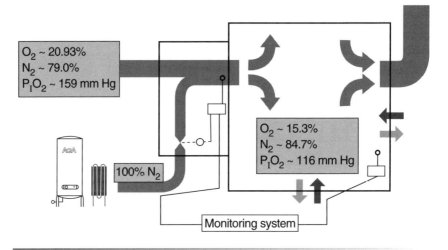

Figure 6.1 Schematic representation of normobaric hypoxic apartment.
Reprinted, by permission, from R.L. Wilber, 2001, "Current trends in altitude training," *Sports Medicine* 31: 252.

(~760 mm Hg), but the concentration of inspired oxygen within the nitrogen apartment (O_2 ~15.3%) is less than that at sea level or outside the nitrogen apartment (O_2 ~20.93%). Figure 6.1 illustrates how the O_2 level is reduced to 15.3% in a hypoxic apartment. A ventilation system pulls in ambient air, which is composed of approximately 20.93% O_2 and 79.0% N_2. A gas composed of 100% N_2 is simultaneously introduced into the ventilation system, which results in an internal gas composition of approximately 15.3% O_2 and 84.7% N_2. This normobaric hypoxic environment simulates an altitude of about 2,500 m (8,200 ft).

This section summarizes the scientific literature pertinent to the use of a normobaric hypoxic apartment for the purpose of simulating an altitude environment. Several of these investigations have been conducted at the Research Institute for Olympic Sport in Jyvaskyla, Finland, whereas two studies were done at the Karolinska Institute in Stockholm, Sweden. More recently, the Australian Institute of Sport has published a series of papers regarding the efficacy of hypoxic apartments on red blood cell production. All of these studies used LHTL altitude training, which required athletes to live and sleep in a simulated altitude environment in the hypoxic apartment but to complete their training in sea level (or approximate sea level) conditions. Additional characteristics of these investigations are as follows:

- Subjects:
    - Were well-trained or elite endurance athletes
    - Included members of the Finnish and Australian national teams

- A sea level control group used in 10 of the 12 studies
- Simulated altitude exposure:
  - Living at 2,000 m to 3,000 m (6,560-9,840 ft), 8 to 18 hr per day for 5 to 28 days
  - Training at sea level (7 studies) or 600 m/1,970 ft (5 studies)
- Serum erythropoietin (EPO) and/or reticulocytes:
  - Measured within 5 days after exposure to simulated altitude (exception: Laitinen et al. [1995] on day 15)
- Red blood cell (RBC) mass and/or hemoglobin:
  - Measured within 1 to 7 days postaltitude
- Maximal oxygen consumption ($\dot{V}O_2$max) (4 studies):
  - Measured within 1 to 7 days postaltitude
- Postaltitude sea level performance (2 studies):
  - Measured within 1 to 7 days postaltitude
  - Included a 40-km cycling time trial and a 400-m running time trial

Investigations Conducted in Finland

The initial investigation on the efficacy of the hypoxic apartment was conducted by Rusko et al. (1995). Six female elite cross-country skiers who were members of the Finnish national team completed an 11-day training block in which they lived 14 hr a day at a simulated altitude of 2,500 m/8,200 ft (normobaric hypoxia) and lived and completed their workouts at sea level (ambient normobaric normoxia) during the other 10 hr of the day. Subjects completed their normal training; specific details of the training sessions were not reported. Blood tests conducted on the fourth day of LHTL indicated that serum EPO (31%) and reticulocyte count (50%) were significantly higher versus prealtitude levels. Hemoglobin and hematocrit data were not reported. Thus, this study suggested that an 11-day period of LHTL altitude training utilizing a normobaric hypoxic apartment produced a significant increase in serum EPO and reticulocyte count in elite cross-country skiers. However, this conclusion was open to debate because of the lack of a fitness-matched sea level control group.

Laitinen et al. (1995) evaluated the effects of living and sleeping in a normobaric hypoxic environment in seven male trained runners. This investigation differed from the previous study of Rusko et al. (1995) in that (1) the subjects completed a longer daily altitude exposure; (2) the subjects completed a longer total altitude exposure; and (3) the experimental design included a control group. The athletes completed a 20- to 28-day training block in which they lived 16 to 18 hr a day at a simulated altitude of 2,500 m/8,200 ft (normobaric hypoxia), and lived and completed their workouts at sea level (ambient normobaric normoxia) during the other 6 to 8 hr of the day. A group of fitness-matched subjects performed similar training at sea level. Details of the training sessions were not reported. Blood tests conducted on the 15th day after the athletes began the LHTL period indicated significant increments in serum EPO (84%) and RBC mass (7%), whereas no changes were observed in the sea level control group.

The relationship between changes in serum EPO, RBC mass, and $\dot{V}O_2$max resulting from exposure to normobaric hypoxia was recently investigated by Rusko et al. (1999). Twelve female and male cross-country skiers and triathletes lived 12 to 16 hr per day for 25 days at a simulated altitude of 2,500 m/8,200 ft (normobaric hypoxia) and completed their workouts at sea level (normobaric normoxia). A group of fitness-matched cross-country skiers and triathletes served as a control group that lived and trained at sea level. Details of the training sessions were not reported. Compared with prealtitude values, serum EPO increased by 60% ($p < 0.05$) in the altitude group when measured on the second day of normobaric hypoxic exposure. In addition, there was a significant 5% increase in RBC mass in the altitude group following 25 days of living in normobaric hypoxia. However, the increase in serum EPO was not significantly correlated to the increase in RBC mass. Sea level $\dot{V}O_2$max was only 1% higher (no statistically significant difference; NSD) in the altitude group on the first day postaltitude, but was significantly higher (3%) on the seventh day postaltitude. The authors concluded that living at a simulated altitude of 2,500 m (8,200 ft) for 25 days produces significant increases in serum EPO and RBC mass, which in turn lead to an improvement in sea level $\dot{V}O_2$max that is evident approximately one week postaltitude.

The response of athletes exposed to a simulated altitude higher than 2,500 m (8,200 ft) was investigated by Mattila and Rusko (1996). Five male competitive cyclists completed an 11-day training block in which they lived 18 hr a day at a simulated altitude of 3,000 m/9,840 ft (normobaric hypoxia) and completed two separate workouts at sea level (ambient normobaric normoxia) during the other 6 hr of the day. Details of the training sessions were not reported. Blood tests conducted on the fifth day after the athletes began the altitude training period indicated significant increments in serum EPO (47%) and reticulocyte count (98%). The increase in serum EPO was 52% higher than the increment in serum EPO reported in the initial study of Rusko et al. (1995), which was conducted at a simulated altitude of 2,500 m (8,200 ft) for an equivalent duration (11 days). Similarly, the increase in reticulocyte count reported by Mattila and Rusko (1996) was 96% higher than that reported in the initial study of Rusko et al. (1995) for athletes who lived and slept at a simulated altitude of 2,500 m (8,200 ft). Collectively, these data suggested a potential dose response for serum EPO and reticulocytes due to the increase in simulated altitude from 2,500 m (8,200 ft) to 3,000 m (9,840 ft). Another unique characteristic of the investigation of Mattila and Rusko (1996) was the fact that it included evaluation of athletic performance (40-km cycling time trial). Compared with prealtitude values, 40-km time trial performance was 4% faster ($p < 0.05$) when assessed on the fifth day following the 11-day LHTL period. Although these data supported the use of a hypoxic apartment for the enhancement of hematological variables and sea level endurance performance, they must be interpreted with caution since there was no sea level control group.

Nummela and Rusko (2000) recently investigated the effect of living in a normobaric hypoxic environment on anaerobic performance. Eight female and male 400-m sprinters from the Finnish national team completed a 10-day training block in which they lived 16 to 17 hr a day at a simulated altitude of 2,200 m/7,215 ft (normobaric hypoxia) but completed their workouts at sea level (ambient normobaric normoxia). Training volume was relatively low during the 10-day experimental period; emphasis was on the development of running speed and speed endurance. A group of 10 male 400-m sprinters completed a similar program at sea level. The athletes were evaluated within seven days after completion of the LHTL training block. Compared with prealtitude values, 400-m time trial performance was 1% faster ($p < 0.05$) in the altitude-trained sprinters. This improvement was equivalent to approximately 0.4 sec. In contrast, there was no difference in pre- versus posttraining 400-m run time for the sea level group. Results of a maximal anaerobic run test (MART) indicated that the LHTL group ran significantly faster than the control group at 5.0 mmol $\cdot$ L^{-1} (running velocity at 5.0 mmol $\cdot$ L^{-1} [$V_{5mmol/L}$]) and 7.0 mmol $\cdot$ L^{-1} ($V_{7mmol/L}$) blood lactate concentrations; these indexes have been shown to be strong predictors of anaerobic performance in track athletes (Nummela et al. 1996a, 1996b). The authors speculated that the improvement in 400-m sprint time seen in the altitude-trained group may have been due to enhanced buffering capacity as evidenced by the fact that resting blood pH was significantly higher in the LHTL sprinters (six of the eight athletes showed an increase in resting blood pH) versus the control sprinters. Thus, the data of this well-controlled study suggested that anaerobic performance in elite 400-m sprinters is enhanced as a result of 10 days of LHTL altitude training.

Investigations Conducted in Sweden

Investigations conducted in Sweden are unique in that they included evaluation of the effect of normobaric hypoxia on postaltitude hemoglobin and hematocrit, in addition to measurement of serum EPO and reticulocyte response. Piehl-Aulin et al. (1998) evaluated six female and male endurance athletes who completed a 10-day training block in which they lived 12 hr a day at a simulated altitude of 2,000 m/6,560 ft (normobaric hypoxia). The subjects lived and completed workouts at sea level (ambient normobaric normoxia) during the other 12 hr of the day. They performed normal sea level training, although specific details of the training sessions were not reported. A control group of five endurance athletes lived and trained at sea level during the 10-day training period. Blood tests conducted within the initial two days of LHTL indicated that serum EPO (80%) and reticulocyte count (60%) were significantly higher versus prealtitude levels. These increments in serum EPO and reticulocytes were similar to those reported in investigations conducted in Finland (Laitinen et al. 1995; Mattila and Rusko 1996; Rusko et al. 1995). Postaltitude measurements undertaken one week after completion of training indicated that hemoglobin (–2%) and hematocrit (–3%) were slightly, but not significantly, lower compared with baseline values. These postaltitude decrements in hemoglobin and hematocrit were attributed to hemodilution. In addition, there was no significant difference in postaltitude $\dot{V}O_2$max versus prealtitude values. Thus, this controlled study suggested that a 10-day period of LHTL altitude training produces a significant increase in serum EPO but does not significantly alter postaltitude hemoglobin concentration, hematocrit, or $\dot{V}O_2$max in trained endurance athletes.

Additional work was conducted by Piehl-Aulin et al. (1998) at a higher simulated altitude. Nine female and male endurance athletes completed a 10-day training block in which they lived 12 hr a day at a simulated altitude of 2,700 m/8,855 ft (normobaric hypoxia) and lived and performed workouts at sea level (ambient normobaric normoxia) during the other 12 hr of the day. Athletes completed their normal training; specific details of the training sessions were not reported. A control group of five endurance athletes lived and trained at sea level during the 10-day training period. Blood tests conducted within the initial five days of LHTL indicated that serum EPO (85%) and reticulocyte count (38%) were significantly higher versus prealtitude levels. Postaltitude measurements undertaken one week after completion of training indicated that hemoglobin was higher (3%, NSD) and hematocrit was lower (–4%, NSD) compared with baseline values. Further, there was no significant difference in postaltitude $\dot{V}O_2$max versus prealtitude values for the LHTL group. Collectively, these data suggested that a 10-day period of LHTL altitude training produces a significant increase in serum EPO but does not affect postaltitude hemoglobin concentration, hematocrit, or $\dot{V}O_2$max in trained endurance athletes.

Investigations Conducted in Australia

Researchers affiliated with the Australian Institute of Sport have also published data regarding the efficacy of using a hypoxic apartment for the enhancement of RBC production. Ashenden et al. (1999b) investigated six female road cyclists from the Australian national team who slept for 12 nights at a simulated altitude of 2,650 m/8,690 ft (normobaric hypoxia) and trained at 600 m/1,970 ft (ambient normobaric normoxia). Six teammates from the Australian national cycling team served as a control group that lived and trained at 600 m/1,970 ft. For both groups, weekly training distance ranged from 515 km to 750 km (319-465 miles) and included two to three interval training sessions. Approximately 20% of the total training time involved an exercise intensity greater than 85% of maximum heart rate. Reticulocyte parameters that have been shown to be sensitive to changes in erythropoiesis (percent reticulocytes, mean corpuscular hemoglobin concentration, reticulocyte hemoglobin) were measured 21 days and 1 day before simulated altitude exposure, after 7 nights and 12 nights of simulated altitude exposure, and again 15 days and 33 days after athletes left the hypoxic apartment. Total hemoglobin mass was measured according to a similar schedule using the carbon monoxide (CO) rebreathing technique. There were no significant changes in any of the reticulocyte parameters between the experimental groups at any time point. Similarly, total hemoglobin mass did not change significantly over the duration of the study, and there were no differences between the two groups. The authors concluded that in elite female road cyclists, 12 nights of exposure to a simulated altitude of 2,650 m (8,690 ft) was not sufficient to stimulate reticulocyte production or increase hemoglobin mass.

Ashenden et al. (2000) recently investigated the effect of intermittent exposure to simulated altitude. Six well-trained male middle distance runners spent five nights (8 to 11 hr per night) in a hypoxic apartment at a simulated altitude of 2,650 m/8,690 ft (normobaric hypoxia) and trained at 600 m/1,970 ft (ambient normobaric normoxia). The five-night period at simulated altitude was followed by three days of living and training at 600 m (1,970 ft). This eight-day period of sleeping five nights at simulated altitude followed by living three days at 600 m was repeated on two subsequent occasions, with all three eight-day sessions completed within 24 days. Five teammates served as a control group, living and training at 600 m (1,970 ft) throughout the duration of the study. The two groups followed an identical training schedule conducted at 600 m, which consisted of endurance runs and track interval workouts. Serum EPO was measured after one and five nights during each of the five-night altitude sessions. Results indicated that serum EPO was significantly higher in the altitude group relative to baseline after one night (57%) and five nights (42%) of sleeping in the hypoxic apartment during the first five-night session. Interestingly, the change in serum EPO was not as pronounced in the second five-night session (first night = 13%; fifth night = –4%) and the third five-night session (first night = 26%; fifth night = 14%) compared with baseline values, suggesting a down-regulation of the hypoxia-induced EPO response. Despite the significant increase in serum EPO observed in the first five-night altitude session, there were no changes in several reticulocyte measures (percent reticulocytes, mean corpuscular hemoglobin concentration, reticulocyte hemoglobin). Indeed, there were no significant increments in reticulocyte parameters or hemoglobin for either group at any point in the study. Ashenden et al. (2000) concluded that brief intermittent exposures to a simulated altitude of 2,650 m (8,690 ft) produced a significant increase in serum EPO; however, this increase in serum EPO was attenuated after the first five-day exposure and was insufficient to stimulate reticulocyte production in trained male distance runners.

In light of their other findings, Ashenden et al. (1999a) conducted an investigation in which the hypoxic stimulus was increased by having athletes live at a higher simulated altitude for a longer period of time as compared with the conditions in the other studies (Ashenden et al. 1999b, 2000). Six male endurance athletes (cyclists, triathletes, cross-country skiers) completed a 23-day training block in which they spent 8 to 10 hr per night at a simulated altitude of 3,000 m/9,840 ft (normobaric hypoxia) and performed workouts at 600 m/1,970 ft (ambient normobaric normoxia). Seven fitness-matched athletes served as a control group. Each athlete kept a daily log of the mode, frequency, intensity, and duration of his training. Over the 30 days of the investigation, the experimental group completed an average of 7.1 training sessions per week, whereas the control group completed 6.8 workouts per week. Total hemoglobin mass, measured using the CO rebreathing technique, was not significantly different between the experimental groups at baseline (LHTL = 990, control = 1,042 g) or after 23 days of simulated altitude exposure (LHTL = 972, control = 1,033 g). Similarly, there were no significant changes between the experimental groups at any time for several reticulocyte parameters.

Ashenden et al. (1999a) speculated that the discrepancy between their results and those reported in previous studies conducted in Finland and Sweden may have been due to methodological differences in the assessment of total hemoglobin mass. Two of the investigations conducted in Finland (Laitinen et al. 1995; Rusko et al. 1999), in which hemoglobin mass was assessed using the Evans blue dye technique, indicated significant increases in total hemoglobin mass following simulated altitude exposure; in contrast, in Australian studies (Ashenden et al.

1999a, 1999b, 2000), hemoglobin mass was assessed using the CO rebreathing technique and showed no significant changes. Ashenden et al. (1999a) further suggested that the Evans blue dye technique may detect a spurious increase in total hemoglobin and RBC mass due to increased albumin leakage from the vasculature resulting from altitude exposure (Poulsen et al. 1998).

Additional studies conducted in Australia have evaluated the efficacy of hypoxic apartments on non-hematological parameters. A recent study by Martin et al. (2002) assessed the effect of LHTL on five Australian National Team female cyclists who slept for 12 nights at a simulated altitude of 2,650 m (8,690 ft) in a nitrogen-diluted apartment and trained at 600 m (1,970 ft). Compared with a performance-matched group of teammates who slept and trained at 600 m, the LHTL cyclists increased their maximal mean power output during a 4-min cycle ergometer time trial (MMP_{4min}) by 2.3% compared with a 0.1% increase for the sea level cyclists. However, the sea level cyclists increased their maximal mean power output during a 30-min cycle ergometer time trial (MMP_{30min}) by 2.4%, whereas the LHTL cyclists decreased by 1.1%. Gore et al. (2001) evaluated six male triathletes, cyclists, and cross-country skiers who slept for 23 nights at a simulated altitude of 3,000 m/9,840 ft (normobaric hypoxia) and completed their training at 600 m/1,970 ft (normobaric normoxia). A fitness-matched control group of seven endurance athletes lived and trained at 600 m (1,970 ft). The LHTL group completed 7.1 training sessions per week (equivalent to 13.4 hr per week), whereas the control group completed 6.8 training sessions per week (equivalent to 10.6 hr per week). Physiological testing was conducted five days before exposure to normobaric hypoxia (pre), after 10 nights of exposure (mid), and two days after 23 nights of exposure (post). For the normobaric hypoxic group, submaximal oxygen consumption ($L \cdot min^{-1}$) at 36%, 52%, 68%, and 84% $\dot{V}O_2$peak was 4% lower ($p < 0.05$) and mechanical efficiency was 1% better ($p < 0.05$) following the 23-day LHTL period. In addition, skeletal muscle buffer capacity increased significantly (18%) in the normobaric hypoxic group after the 23-day LHTL period. The precise mechanisms responsible for enhanced skeletal muscle buffer capacity following altitude training are unclear but may be related to changes in creatine phosphate, muscle protein concentrations, or both (Mizuno et al. 1990). There were no changes over time in any of the physiological measures for the control group.

Roberts et al. (2003) recently reported on the difference between 5, 10, and 15 days of LHTL. Nineteen trained cyclists (14 males, 5 females, average $\dot{V}O_2$max = 62.3 ml $\cdot$ kg^{-1} $\cdot$ min^{-1}) were divided into three groups and randomly assigned to complete a period of either 5, 10, or 15 days. During that time they spent 8 to 10 hr per night at a simulated altitude of

2,650 m (8,690 ft) in a nitrogen apartment and trained at 600 m (1,970 ft). The athletes also completed a corresponding period of 5, 10, or 15 days in a control condition—that is, they lived and trained at 600 m during the entire day. For LHTL, there were no differences in any physiological or performance parameters between the three durations, suggesting that 10 or 15 days of LHTL is not more effective than 5 days. However, when the data from all three LHTL groups were combined, $MMPO_{4min}$ and maximal accumulated oxygen deficit (MAOD) were significantly improved by 4% and 10%, respectively, following LHTL compared with pre-LHTL. $MMPO_{4min}$ and MAOD were not changed during the control condition in these athletes. These data suggested that LHTL may enhance performance in middle distance events—for example, the 4,000-m team pursuit cycling event or the 1,500-m run.

The effect of continuous versus intermittent LHTL protocols on the hypoxic ventilatory response (HVR) was reported recently by Townsend and colleagues (2002). Recall from chapter 2 that the HVR is a noninvasive method used to evaluate the sensitivity of the peripheral chemoreceptors to hypoxia. In this study, 33 trained endurance athletes were divided into three groups. One group (LHTLc) spent 20 nights, 8 to 10 hr per night, at a simulated elevation of 2,650 m (8,690 ft) in a nitrogen apartment and trained at 600 m (1,970 ft). Another group (LHTLi) followed the same protocol as LHTLc except the 20-night period was divided into four 5-night blocks, each separated by 2 nights of normoxia. A third group that lived and trained at 600 m served as a control group. The main findings of this study were: (1) compared to the control group, HVR increased significantly during LHTLi and LHTLc, with the overall response being more evident during consecutive nightly exposure versus intermittent nightly exposure; and (2) individual increases in HVR were variable, with stronger individual responses tending to occur during consecutive nightly exposure (Townsend et al. 2002).

Summary

Currently, elite athletes in several countries utilize normobaric hypoxic apartments for the purpose of simulating an altitude living environment. Typically they live and sleep high in a simulated altitude environment in the hypoxic apartment for 8 to 18 hr a day but complete their training in sea level or approximate sea level conditions. Data from several of the studies reviewed in this section suggest that using a hypoxic apartment in this manner may produce beneficial changes in serum EPO, reticulocyte count, and RBC mass, which in turn may lead to improvements in postaltitude endurance performance. However, other studies have failed to demonstrate significant changes in erythropoietic indexes as a result of normobaric hypoxic exposure. The discrepancy in these findings may be attributable in part to methodological differences used to assess total hemoglobin and RBC mass (i.e., Evans blue dye technique vs. CO rebreathing technique). In addition, differences in the hypoxic stimulus that the subjects were exposed to (i.e., elevation and duration of exposure), or the training status of the subjects (trained vs. elite national team athletes), or both, may account for the inconsistent results. A limited number of studies have suggested that anaerobic capacity and performance are enhanced as a result of use of a hypoxic apartment. Table 6.1 summarizes the key findings of the scientific literature pertinent to altitude training used in conjunction with a normobaric hypoxic apartment.

Supplemental Oxygen

Several studies have evaluated the efficacy of supplemental oxygen on $\dot{V}O_2$max and exercise performance in athletes at sea level. Powers et al. (1989) reported that mild hyperoxia (fraction of inspired oxygen [F_IO_2] ~0.26) increased $\dot{V}O_2$max in highly trained athletes whose arterial oxyhemoglobin saturation (S_aO_2) decreased to less than 92% during maximal exercise in normoxic conditions; $\dot{V}O_2$max was unaffected by supplemental oxygen in athletes who did not develop hypoxemia during maximal exercise in normoxic conditions. Additional investigations have documented the beneficial effects of acute hyperoxia (F_IO_2 ~0.30-0.62) on physiological responses and performance during continuous and intermittent exercise in well-trained rowers (Peltonen et al. 1995), endurance athletes (Peltonen et al. 1999), and sprinters (Nummela et al. 2000). Hamalainen et al. (2000) recently reported that a four-week training program in which trained runners used supplemental oxygen for three workouts per week produced significant improvements in sea level 3,000-m running performance (3%) and maximal anaerobic running velocity (3%). Submaximal and maximal exercise using supplemental oxygen (F_IO_2 ~0.30-0.70) has been associated with a reduced rate of lactate accumulation in muscle and blood (Graham et al. 1987; Howley et al. 1983); attenuated arterial oxyhemoglobin desaturation (Peltonen et al. 1999); maintenance of resting levels of adenosine triphosphate (ATP) and adenosine diphosphate (ADP), as well as total reduced nicotinamide adenine dinucleotide (NADH) (Linossier et al. 2000); and enhanced peak power output (Hughson and Kowalchuk 1995). Additional information on hyperoxia and its effect on sea level endurance performance is provided in a review by Welch (1987).

Supplemental oxygen has also been used for the purpose of simulating either normoxic (i.e., sea level) or hyperoxic conditions during high-intensity workouts conducted at altitude. Use of supplemental oxygen

Table 6.1 Scientific Literature About the Use of a Normobaric Hypoxic Apartment and LHTL Altitude Training

Author	Altitude (m/ft)	Subjects	SL control	Altitude exposure (days)	Δ EPO[a] (%)	Δ Reticulocytes[a] (%)	Δ RCM[b] (%)	Δ Hb[b] (%)	Δ $\dot{V}O_2$max[b] (%)	Post-altitude SL performance[b]
Piehl-Aulin et al. 1998	Live: 2,000/ 6,560 (12 hr · day^{-1}) Train: SL	Female and male endurance athletes (n = 6)	yes	10	80*	60*	NR	−2 NSD	NSD	NR
Nummela and Rusko 2000	Live: 2,200/ 7,215 (16-17 hr · day^{-1}) Train: SL	Female and male 400-m runners (n = 8) Finnish NT	Yes	10	NR	NR	NR	NR	NR	400-m TT faster by 1% (0.4 sec)*
Laitinen et al. 1995	Live: 2,500/ 8,200 (16-18 hr · day^{-1}) Train: SL	Male trained runners (n = 7)	Yes	20-28	84*	NR	7*	NR	NR	NR
Rusko et al. 1995	Live: 2,500/ 8,200 (14 hr · day^{-1}) Train: SL	Female elite CC skiers (n = 6) Finnish NT	No	11	31*	50*	NR	NR	NR	NR
Rusko et al. 1999	Live: 2,500/ 8,200 (12-16 hr · day^{-1}) Train: SL	Female and male endurance athletes (n = 12)	yes	25	60*	NR	5*	NR	3*	NR

Study	Altitude	Subjects		Duration						Results
Ashenden et al. 1999b	Live: 2,650/8,690 (8-10 hr · day⁻¹) Train: 600/1,970	Female elite cyclists (n = 6) Australian NT	Yes	12	NR	~7 NSD	NR	3 NSD	NR	NR
Ashenden et al. 2000	Live: 2,650/8,690 (8-11 hr · day⁻¹) Train: 600/1,970	Male trained runners (n = 6)	Yes	Three 5-day sessions (with a 3-day interval at 600 m between each 5-day session)	57* (after 24 hr of the first 5-day session)	NSD	NR	NSD	NR	NR
Martin et al. 2002	Live: 2,650/8,690 (8-10 hr · day⁻¹) Train: 600/1,970	Female elite cyclists (n = 5) Australian NT	Yes	12	NR	NR	NR	NR	NR	$MMPO_{4\,min}$ increased by 2.3%* $MMPO_{30\,min}$ decreased by 1.1%*
Townsend et al. 2002	Live: 2,650/8,690 (8-10 hr · day⁻¹) Train: 600/1,970	Female and male endurance athletes (n = 33)	Yes	20 days consecutively (LHTLc) or 20 days divided into four 5-night blocks (LHTLi)	NR	NR	NR	NR	NR	HVR increased significantly during LHTLi and LHTLc, with the overall response being more evident during LHTLc
Roberts et al. 2003	Live: 2,650/8,690 (8-10 hr · day⁻¹) Train: 600/1,970	Female and male trained cyclists (n = 19)	Yes	5, 10, or 15	NR	NR	NR	NR	NSD	$MMPO_{4\,min}$ and MAOD increased by 4%* and 10%*, respectively
Piehl-Aulin et al. 1998	Live: 2,700/8,855 (12 hr · day⁻¹) Train: SL	Female and male endurance athletes (n = 9)	Yes	10	85*	38*	NR	3 NSD	NSD	NR

(continued)

Table 6.1 (continued)

Author	Altitude (m/ft)	Subjects	SL control	Altitude exposure (days)	Δ EPO[a] (%)	Δ Reticulocytes[a] (%)	Δ RCM[b] (%)	Δ Hb[b] (%)	Δ $\dot{V}O_2max$[b] (%)	Post-altitude SL performance[b]
Mattila and Rusko 1996	Live: 3,000/9,840 (18 hr · day⁻¹) Train: SL	Competitive male cyclists (n = 5)	No	11	47*	98*	NR	NR	NR	40-km TT faster by 4%*
Gore et al. 2001	Live: 3,000/9,840 (8-10 hr · day⁻¹) Train: 600/1,970	Male endurance athletes (n = 6)	Yes	23	NR	NR	NR	NR	#	NR
Ashenden et al. 1999a	Live: 3,000/9,840 (8-10 hr · day⁻¹) Train: 600/1,970	Male endurance athletes (n = 6)	Yes	23	NR	~18 NSD	NR	−2 NSD	NR	NR

Studies are listed by ascending living altitude.

Δ = change.

[a] Measured within 1 to 5 days after arriving at altitude. Exception: Laitinen et al. (1995) on day 15.

[b] Measured within 1 to 7 days postaltitude.

* Significant difference versus prealtitude (p < 0.05).

Submaximal $\dot{V}O_2$ (L · min⁻¹) at workloads equivalent to 36%, 52%, 68%, and 84% $\dot{V}O_2$peak were significantly lower postaltitude versus prealtitude (p < 0.05).

CC = cross-country; EPO = erythropoietin (mU · ml⁻¹); Hb = hemoglobin (g · dL⁻¹); HVR = hypoxic ventilatory response; LHTL = "live high-train low"; MAOD = maximal accumulated oxygen deficit; MMPO$_{4 min}$ = maximal mean power output during a 4-min cycle ergometer time trial; MMPO$_{30min}$ = maximal mean power output during a 30-min cycle ergometer time trial; NR = not reported/measured; NSD = no statistically significant difference; NT = national team; RCM = red cell mass (ml · kg⁻¹); SL = sea level; TT = time trial; $\dot{V}O_2$max = maximal oxygen consumption.

in this manner is essentially a modification of the LHTL strategy in that athletes live in a natural, hypobaric hypoxic altitude environment but train at "sea level" with the aid of supplemental oxygen. This system is used effectively at the U.S. Olympic Training Center in Colorado Springs, Colorado, where U.S. national team athletes live at 1,860 m (6,100 ft) or higher but can train at "sea level" with the aid of supplemental oxygen (figure 6.2). The average barometric pressure in Colorado

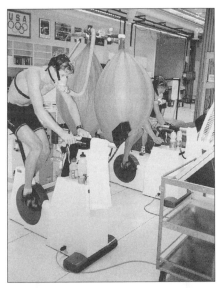

Figure 6.2 Supplemental oxygen training by U.S. national team triathletes at the United States Olympic Training Center in Colorado Springs, Colorado (1,860 m/6,100 ft).

Courtesy of the author.

Springs is approximately 610 mm Hg, which yields a partial pressure of inspired oxygen (P_IO_2) of approximately 128 mm Hg. By inhaling a certified medical gas composed of 26.5% oxygen, athletes can complete high-intensity training sessions in a "sea level" environment at a P_IO_2 equivalent to 159 mm Hg (760 mm Hg; F_IO_2 ~0.2093) (figure 6.3).

Only a few studies have evaluated the efficacy of using supplemental oxygen in conjunction with LHTL altitude training (LH + TLO_2). Wilber et al. (2003b) recently reported on the acute effects of supplemental oxygen on physiological responses and exercise performance during a high-intensity interval workout in trained endurance athletes living at altitude (1,800-1,900 m/5,905-6,230 ft). Testing was conducted at the U.S. Olympic Training Center in Colorado Springs, Colorado, at an elevation of 1,860 m/6,100 ft (barometric pressure 610-612 mm Hg, P_IO_2 ~128 mm Hg). The athletes completed three randomized trials in which they performed a standardized interval workout while inspiring a medical grade gas with an oxygen concentration of approximately 21% (P_IO_2 ~128 mm Hg), 26% (P_IO_2 ~159 mm Hg), and 60% (P_IO_2 ~366 mm Hg). The standardized interval workout consisted of 6 × 100 kilojoules (kJ) of work performed on a cycle ergometer at a self-selected workload and pedaling cadence with a work:recovery ratio of 1:1.5. Compared with the control trial (21% O_2), average total time (min:sec) for the 100-kJ

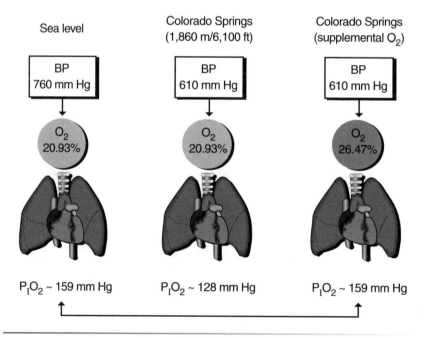

Figure 6.3 Scientific rationale for using supplemental oxygen for the purpose of simulating sea level P_IO_2 at moderate altitude (1,860 m/6,100 ft).

work interval was 5% and 8% (p < 0.05) faster in the 26% O_2 and 60% O_2 trials, respectively (figure 6.4a). Consistent with the improvements in total time were increments in power output (W) equivalent to 5% in the 26% O_2 trial and 9% in the 60% O_2 trial (p < 0.05) (figure 6.4b). Whole-body $\dot{V}O_2$ (L · min^{-1}) was higher by 7% and 14% (p < 0.05) in the 26% O_2 and 60% O_2 trials, respectively (figure 6.4c), and was highly correlated with the improvement in power output (r = 0.85, p < 0.05). Arterial oxyhemoglobin saturation (S_pO_2%) was significantly higher by 5% (26% O_2) and 8% (60% O_2) in the supplemental oxygen trials and demonstrated a minor but significant correlation with $\dot{V}O_2$ (r = 0.28, p < 0.05) (figure 6.4d). In addition, there were no significant differences between the supplemental oxygen trials and the control trial in serum lipid hydroperoxide (LOOH) and reduced glutathione (GSH) or urinary malondialdehyde (MDA) and 8-hydroxy-deoxygenase (8-OHdG), suggesting that there is no additional oxidative stress during supplemental oxygen workouts (Wilber et al. 2003a). The authors concluded that LH + TLO$_2$ altitude training results in significant increases in arterial oxyhemoglobin saturation, $\dot{V}O_2$, and average power output contributing to a significant improvement

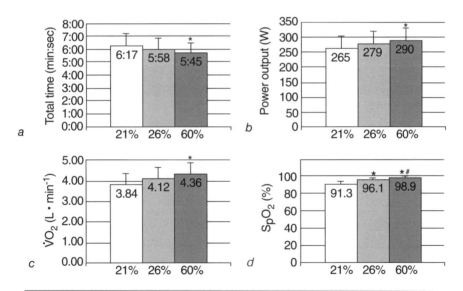

Figure 6.4 Effect of supplemental oxygen (26.0% O_2, 60.0% O_2) on (a) average total time, (b) power output, (c) oxygen uptake ($\dot{V}O_2$), and (d) oxyhemoglobin saturation in trained male cyclists during an interval training workout conducted at 1,860 m (6,100 ft). The interval workout consisted of 6 × 100 kJ using a work-to-rest ratio of 1:1.5. * Significantly different versus ambient air (20.93% O_2) (p < 0.05). #Significantly different versus 26% O_2 (p < 0.05).

Reprinted, by permission, from R. Wilber et al., 2003, "Effect of F$_I$O$_2$ on physiological responses and cycling performance at moderate altitude," *Medicine and Science in Sports and Exercise* 35: 1155-1156.

in exercise performance without inducing additional oxidative stress. In terms of practical application, these data provided support for using supplemental oxygen as an altitude training strategy that allows athletes to live high at moderate altitude and effectively "train low" with minimal travel and inconvenience.

A few studies have looked at the long-term training effects of using supplemental oxygen in conjunction with LHTL altitude training. Chick et al. (1993) evaluated the effect of supplemental oxygen training on five trained female and male endurance athletes, who lived and trained at an elevation of 1,600 m/5,250 ft (Albuquerque, NM). The athletes completed baseline training until they were "maximally trained," as defined by three consecutive laboratory tests in which maximal cycling time did not increase by more than 2% and endurance time at 85% maximal workload (ET85) did not increase by more than 8% compared with the best result of previous tests. Once the athletes were maximally trained at 1,600 m (5,250 ft), they began six weeks of progressive cycle training using supplemental oxygen (F_IO_2 ~0.70). Training consisted of 8 × 5-min sets at 95% maximal workload done four days per week for six weeks. The athletes could not do this workout for more than 10 min in normoxic conditions at 1,600 m, but completed the entire workout using supplemental oxygen. Following the 42-day supplemental oxygen training period, the athletes were evaluated at 1,600 m (5,250 ft). Compared with values obtained before hyperoxic training, endurance time during a maximal cycling test increased from 19:06 (min:sec) to 19:36 ($p < 0.05$); ET85 heart rate decreased from 168 to 163 bpm ($p < 0.05$); and ET85 increased from 6:12 to 8:12 ($p < 0.05$). These data suggested that supplemental oxygen improves an athlete's ability to perform high-intensity workouts at altitude. However, the findings are open to debate because of the lack of a normoxic-trained control group.

Morris et al. (2000) also evaluated the long-term training effects of using supplemental oxygen during high-intensity workouts at altitude (1,860 m/6,100 ft). Eight male junior cyclists who were members of the U.S. national team completed a 21-day training period during which they lived and performed their base workouts at 1,860 m/6,100 ft (Colorado Springs, CO) but performed their interval training in simulated "sea level" conditions using a gas mixture consisting of approximately 26% oxygen (P_IO_2 ~159 mm Hg). Interval workouts were done three days per week and required the athletes to complete 5 × 5-min cycling efforts at 105% to 110% of maximal steady-state heart rate. A control group of eight male cyclists completed the same training program at 1,860 m using a normoxic gas mixture (F_IO_2 ~0.2093, P_IO_2 ~128 mm Hg). Athletes using supplemental oxygen were able to train at a significantly higher percentage of their altitude lactate threshold (126%) than their counter-

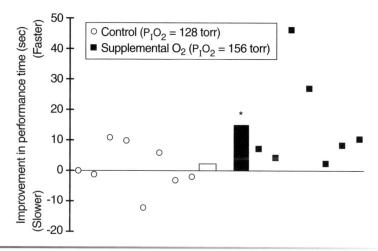

Figure 6.5 Individual and group improvement in the performance of a 120-kJ cycling time trial. * Significantly different versus pretraining value for same group (p < 0.05).

Reprinted, by permission, from D.M. Morris et al., 2000, "The effects of breathing supplemental oxygen during altitude training on cycling performance," *Journal of Science and Medicine in Sport* 3: 172.

parts who trained in normoxic conditions (109%). Following the 21-day training period, the athletes performed a 120-kJ cycling performance time trial in simulated sea level conditions (F_IO_2 ~0.26, P_IO_2 ~159 mm Hg). Results of the cycling performance test showed improvements of 2 and 15 sec for the normoxic-trained and hyperoxic-trained cyclists, respectively. The 15-sec improvement in cycling performance seen in the hyperoxic-trained group was significant ($p<0.05$) compared to their pretraining performance (figure 6.5). With results similar to those of Chick et al. (1993), Morris et al. (2000) demonstrated that high-intensity workouts at moderate altitude (1,860 m/6,100 ft) are enhanced through the use of supplemental oxygen that simulates sea level conditions. In addition, Morris et al. (2000) showed that sea level endurance performance is enhanced when supplemental oxygen is used in conjunction with LHTL altitude training.

Table 6.2 provides an example of a supplemental oxygen workout used by U.S. national team pursuit cyclists living at moderate altitude (1,860 m/6,100 ft) at the U.S. Olympic Training Center in Colorado Springs. The workout consists of a series of 15-sec efforts done at a minimum pedaling cadence of 140 rpm. The first 15-sec effort is done at a power output (W) equivalent to 110% of the power required of a cyclist in the front position (position 1) on an internationally competitive pursuit cycling team. The second effort follows immediately and requires the athlete to cycle for 15 sec at a power output equivalent to 110% of the power required of a cyclist in the back position (position 4) on an internationally competitive pursuit cycling team. Three more 15-sec efforts are done at the power output for positions 1, 4, and 1, respectively. The athlete then recovers actively for 5 min without supplemental oxygen. Following recovery, the athlete repeats the entire 15-sec sequence six times. Table 6.3 provides some representative data for U.S. national team pursuit cyclists who have used this supplemental oxygen workout while living year-round at altitude. Average power output and peak power output during the "position 1" efforts are approximately 50% and 60% greater, respectively, than the athlete's power output at $\dot{V}O_2$max (475-500 W; measured at 1,860 m/6,100 ft).

Table 6.4 presents an example of a supplemental oxygen workout used by U.S. national team triathletes living at moderate altitude at the U.S. Olympic Training Center in Colorado Springs. The workout consists of three 10-min runs done at progressively faster paces (4:56, 4:54, and 4:52 mile pace). A 4-min recovery jog without supplemental oxygen follows each of the 10-min work intervals. Table 6.5 shows the sequence and progression of several supplemental oxygen workouts done over the final four months of the season by a male U.S. Olympic team triathlete; it also provides the athlete's comments relative to each workout. In general, most of the supplemental oxygen training consisted of relatively long intervals (10-30 min) done slightly below or at maximal steady-state running pace. These longer, moderate-intensity workouts were used in order to decrease the possibility of injury or overtraining.

▌ Table 6.2 Example of a Supplemental Oxygen Workout for Elite Cyclists

Warm-up	Workout (minimum pedaling cadence = 140 rpm)	Warm-down
30-45 min	15 sec @ 110% position 1 power	45-60 min
	15 sec @ 110% position 4 power	
	15 sec @ 110% position 1 power	
	15 sec @ 110% position 4 power	
	15 sec @ 110% position 1 power	
	5-min active recovery without supplemental O_2	
	15-sec sequence repeated six times	

Used by U.S. national team pursuit cyclists living at moderate altitude (1,860 m/6,100 ft).

Position 1 power = power in watts (W) required in the front position on an internationally competitive pursuit cycling team.

Position 4 power = power in watts (W) required in the back position on an internationally competitive pursuit cycling team.

Table 6.3 Data for U.S. National Team Pursuit Cyclists Who Have Used Supplemental Oxygen Training*

Time (sec)	Average cadence (rpm)	Average power (W)	Peak power (W)
0-15	143	727	767
15-30	148	294	326
30-45	136	708	739
45-60	147	292	330
60-75	134	708	732
5-min recovery			
0-15	140	732	768
15-30	148	292	314
30-45	139	730	746
45-60	148	295	311
60-75	139	724	756

Data are from cyclists who have used supplemental oxygen training while living year-round at moderate altitude (1,860 m/6,100 ft).

Average power output and peak power output during the "position 1" efforts are approximately 50% and 60% greater, respectively, than the athlete's power output at $\dot{V}O_2$max (475-500 W; measured at 1,860 m/6,100 ft).

Laboratory test values (measured at 1,860 m/6,100 ft):
Power at blood lactate threshold = 300-350 W
$\dot{V}O_2$max = 65-75 ml · kg^{-1} · min^{-1}
Power at $\dot{V}O_2$max = 475-500 W

* = As shown in table 6.2.
rpm = revolutions per minute; W = watts.

Table 6.4 Example of a Supplemental Oxygen Workout for Elite Triathletes

Warm-up	Workout	Warm-down
30 min	10 min @ 4:56 mile pace 4-min jog recovery without supplemental O_2 10 min @ 4:54 mile pace 4-min jog recovery without supplemental O_2 10 min @ 4:52 mile pace	30-45 min

Used by U.S. national team triathletes living at moderate altitude (1,860 m / 6,100 ft)

Table 6.5 Supplemental Oxygen Workouts Over the Final Four Months of the Season by a U.S. Olympic Team Triathlete

Date	Workout	Athlete's comments
7/20	20 min @ 4:56 mile pace	Pace felt comfortable until last 2 min. Four-mile PR. Felt a "crash" about 15 min after workout (light-headed, dead legs). Legs were very fatigued next day.
7/27	3 × 12 min @ 4:56/4:54/4:52 mile pace with 3-min recovery	First two intervals were comfortable, third interval became difficult in the final 4 min. Experienced post-workout "crash" again. Legs were fatigued for 2 days.
7/30	15 min @ 4:54 mile pace	Very tired to begin with, pace felt difficult from the start. This is my third supplemental O_2 workout in 10 days—maybe a little much? Felt terrible after the workout.
8/9	Wilkes-Barre*	
8/15	Canada ITU World Cup*	
8/19	30 min continuous @ 4:56 (10 min) / 4:54 (10 min) / 4:52 (10 min) mile pace	Felt strong until the last 3 min. Pace felt easy and I felt comfortable. Muscles were not sore, but I felt generally bad for the next 3 days.
8/24	15 min @ 4:56 mile pace 10 min @ 4:54 mile pace 5 min @ 4:52 mile pace 3-min recovery between intervals	I felt great! I was in control. Felt the usual post-workout "crash." Only felt fatigued for 1 day after the workout.
9/12	ITU World Championships*	
9/21	3 × 10 min @ 4:56/4:54/4:52 mile pace with 4-min recovery	Felt the most relaxed of any of my supplemental O_2 workouts. Felt good enough to do the last 3 min of my last interval at 4:50 mile pace. Quickest recovery of any of my supplemental O_2 workouts.
10/10	Cancun ITU World Cup*	

Date	Workout	Athlete's comments
10/13	10 × 3 min @ 4:56/ 4:54/4:52/4:50/4: 50/4:48/4:45/4:43/ 4:41/4:39 mile pace with 2-min recovery	Best workout yet—I could have gone on less than 2-min recovery. 4:39 mile pace felt good as a limit. I felt better the next day than I have after any previous supplemental O_2 workout. My legs were a little bit sore due to the increased speed, but the general fatigue was far less severe.
11/7	Noosa ITU World Cup*	

Athlete was a male U.S. Olympic team triathlete living at moderate altitude (1,860 m / 6,100 ft).

* Competition.

ITU = International Triathlon Union; PR = personal record.

Only near the end of the season were shorter (3 min), faster work intervals attempted. The athlete's comments make it clear that postworkout recovery is an important concern with the use of supplemental oxygen training at altitude. Thus, it is critical that coaches monitor the recovery process very closely in the days following supplemental oxygen workouts in order to avoid overtraining.

The following summarizes some of the key issues that athletes and coaches need to consider when using supplemental oxygen training.

- Athletes are able to work at a higher power output (cycling) and faster velocity/pace (running) than in similar workouts at altitude without supplemental oxygen.
- Ventilation, heart rate, and the athlete's perception of effort are lower than in similar workouts done at altitude without supplemental oxygen.
- Pronounced local muscle fatigue and soreness in the day(s) following supplemental oxygen workouts are common. Substantial additional recovery must be built into the training program to account for this fatigue or soreness.
- It is important to exercise considerable caution to prevent injury or overtraining by providing an adequate warm-up and controlling the rate of workout progression. This is especially true for treadmill workouts.
- Supplemental oxygen training is psychologically tough training. Therefore, supplemental oxygen workouts should be used very selectively.

In summary, supplemental oxygen has been used for the purpose of simulating either normoxic (sea level) or hyperoxic conditions during high-intensity workouts conducted at altitude. Use of supplemental oxygen in this manner is a modification of the LHTL strategy in that athletes live in a natural, hypobaric hypoxic altitude environment but train at "sea level" with the aid of supplemental oxygen. Although limited, scientific data regarding the efficacy of hyperoxic training suggest that high-intensity workouts at moderate altitude (1,860 m/6,100 ft) and endurance performance at sea level may be enhanced through the use of supplemental oxygen.

Hypoxic Sleeping Units

Endurance athletes have recently started to use hypoxic sleeping units as part of their altitude training programs. These include the Colorado Altitude Training (CAT) Hatch™, Hypoxico Altitude Tent™, and Colorado Mountain Room™, all of which are designed to allow athletes to sleep high and train low. The CAT Hatch is a cylindrical hypobaric chamber that allows one person to lie in a supine or prone position; it can simulate altitudes up to approximately 4,575 m (15,005 ft). The CAT Hatch is based on the same technology as the Gamow bag (Gamow et al. 1990), a portable hyperbaric unit that has been used for several years as an emergency medical therapy for mountaineers experiencing acute mountain sickness (AMS), high-altitude cerebral edema, or high-altitude pulmonary edema (Bartsch 1992). The difference, of course, is that the Gamow bag simulates a hyperbaric environment, whereas the CAT Hatch simulates high altitude or a hypobaric hypoxic environment. The CAT Hatch is commercially available at a cost of approximately $15,000 U.S. (2003 price). To date, no scientific studies have been conducted to determine the efficacy of the CAT Hatch on RBC production, $\dot{V}O_2$max, or performance in elite athletes.

Another hypoxic sleeping unit that is currently being used by endurance athletes is the Hypoxico Altitude Tent. This normobaric hypoxic system can be installed over a standard double or queen-sized bed and simulates elevations up to approximately 4,268 m (14,000 ft) (figure 6.6). The Hypoxico Altitude Tent uses an oxygen-filtering membrane that "scrubs" or reduces the concentration of oxygen from the air outside the tent. The oxygen-reduced air is then pumped inside the tent, resulting in a normobaric hypoxic environment. The Colorado Mountain Room uses similar technology but is designed to convert an entire room into a normobaric hypoxic living/sleeping environment. The Hypoxico Altitude Tent is commercially available at a cost of approximately $8,000 U.S. (2003 price), whereas a standard-sized bedroom converted

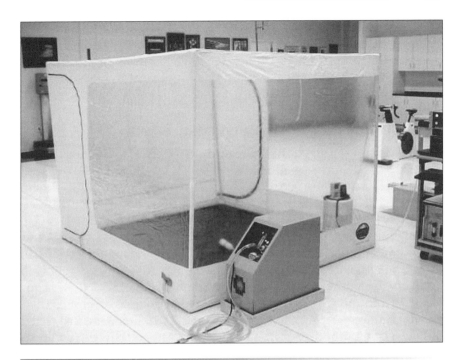

Figure 6.6 The Hypoxico Altitude Tent creates a normobaric hypoxic environment that simulates altitudes up to 4,268 m (14,000 ft).
Courtesy of Deb Whitney (USA Swimming).

to the Colorado Mountain Room costs approximately $15,000 U.S. (2003 price). As with the CAT Hatch, no scientific studies have evaluated the efficacy of the Hypoxico Altitude Tent or Colorado Mountain Room on RBC production, $\dot{V}O_2$max, or performance in elite athletes. However, preliminary research suggests that altitude tents provide a relatively safe and comfortable normobaric hypoxic environment (Shannon et al. 2001; Wilber et al. 2002) but do not significantly alter RBC levels over a four-week period (Ingham et al. 2001).

Intermittent Hypoxic Exposure

The use of intermittent hypoxic exposure (IHE) for the purpose of enhancing athletic performance is based on the fact that brief exposures to hypoxia (1.5-2.0 hr) stimulate the release of EPO (Eckardt et al. 1989; Knaupp et al. 1992; Powell and Garcia 2000; Schmidt 2002). On the basis of these findings, it is assumed that IHE will be sufficient to bring about significant increments in serum EPO and RBC concentration and ultimately enhance $\dot{V}O_2$max and endurance performance. Athletes typically use IHE (1) while at rest or (2) in conjunction with a training session (this use is referred to as intermittent hypoxic training [IHT]). In effect, IHE and IHT allow athletes to "live low-train high." Athletes typically use a hypobaric chamber or a Hypoxicator/Go2Altitude™ device for IHE and IHT sessions.

Intermittent Hypoxic Exposure at Rest

Data are minimal and equivocal regarding the effect of passive IHE on exercise performance in well-trained athletes. Hellemans (1999) reported preliminary data from a study involving 10 elite endurance athletes who completed 20 days of IHE. The athletes breathed hypoxic air through a face mask device (Hypoxicator) using an F_IO_2 of approximately 0.10 (simulating an altitude of ~5,800 m/19,025 ft) for the first 10 days and 0.09 (simulating an altitude of ~6,400 m/20,990 ft) for the second 10 days. The hypoxic exposures were 5 min in length and were followed by equivalent time periods of normoxic exposure. The IHE sessions lasted 60 min and took place twice per day. The athletes continued their normal training during the experimental period, but were instructed to take a minimum of 1 hr of recovery between an IHE session and a workout. Postexperimental results showed significant improvements in endurance performance (3%), reticulocyte count (29%), hemoglobin (4%), and hematocrit (5%). Similarly, Rodriguez et al. (2000) reported that eight trained subjects demonstrated significant increases in reticulocyte percentage (108%), RBC count (7%), hemoglobin (13%), and hematocrit (6%) following passive exposure to hypobaric hypoxia (4,000 m/13,120 ft to 5,500 m/18,040 ft) for 90 min, three times a week, for three weeks. However, no pre- versus postexperimental changes were observed in cycling exercise time or $\dot{V}O_2$max, although power output (W) at the anaerobic threshold was significantly higher following the three-week IHE regimen. It should be noted that a non-IHE control group was not included in the experimental design of either Hellemans (1999) or Rodriguez et al. (2000).

In contrast, Frey et al. (2000) recently reported that passive IHE had no effect on hematological indexes or submaximal or maximal exercise responses in moderately trained females and males. The IHE sessions lasted 75 min and occurred each day for 21 days; the subjects were exposed to an F_IO_2 of approximately 0.09 (simulating an altitude ~6,400 m/20,990 ft) using a Hypoxicator device similar to that employed by Hellemans (1999). Details of the IHE exposure:recovery ratio were not reported. A non-IHE control group was not included in the experimental design. Although there was a significant increase in serum EPO (38%) when measured 2 hr after the first IHE session, there were no other changes in serum EPO, reticulocytes, hemoglobin, or hematocrit during the experimental period. In addition, there were no differences in submaximal $\dot{V}O_2$ at 2.0 and 4.0 mmol · L^{-1} blood lactate concentrations or in $\dot{V}O_2$max. A recent study by Glyde-Julian et al. (2003) reported similar results. National class female and male distance runners used a Hypoxicator/Go2Altitude device to complete an IHE regimen of 5 min "on" + 5 min "off" for 70 min at an altitude of 5,000 m (16,400 ft), five days per week for four weeks. This IHE regimen replicated the manufacturer's recommendations for effective results. Compared with a fitness-matched control group, there were no differences in 3,000-m run time, $\dot{V}O_2$max, serum EPO, or several blood measures of enhanced erythropoiesis following the four weeks of IHE.

Preliminary data suggest that an adequate "dose" of IHE is required in order to effectively stimulate erythropoiesis and enhance aerobic performance in elite endurance track cyclists (Rodriguez et al. 2002a) and well-trained swimmers (Rodriguez et al. 2002b). It appears that this effective dose of IHE requires relatively high levels of hypoxia (>5,000 m/16,400 ft) of sufficient duration (>3 hr per day for two to four weeks) combined with sea level training.

Intermittent Hypoxic Exposure During Exercise

The use of IHE in conjunction with training sessions is referred to as intermittent hypoxic training (IHT). One of the initial studies on the efficacy of IHT was conducted by Terrados et al. (1988). The subjects were eight male elite cyclists who were randomly assigned to two training groups. One group trained in a hypobaric chamber at a simulated altitude of 2,300 m (7,544 ft), whereas the other group trained in a laboratory that was maintained at sea level environmental conditions. Both groups trained for 21 to 28 days, four to five days per week. Training consisted of 60 to 90 min of continuous cycle ergometer exercise at the same relative intensity (65-70% Wmax) followed by 45 to 60 min of interval training. Postexperimental tests were conducted within one to two days of completion of the training period. Compared with preexperimental values, the IHT group showed improvements in hemoglobin (4%) and hematocrit (5%); however, these increments were not statistically different compared with values in the control group. The IHT cyclists demonstrated significant improvements in total work capacity (33%) and maximal power output (12%) during an exhaustive cycling test. The control cyclists also showed improvements in total work capacity and maximal power output; but the increments were not as great as for the IHT cyclists, nor were they statistically significant. Thus, these data suggested that IHT done for 2 to 3 hr a day for 21 to 28 days does not lead to improvements in hemoglobin or $\dot{V}O_2$max compared with equivalent training at sea level, but may enhance performance capabilities (total work capacity, maximal power output) in elite cyclists.

At a similar simulated altitude (2,500 m/8,200 ft), 16 elite triathletes used IHT over a 10-day period for 2 hr a day, completing cycle ergometer workouts in a hypobaric chamber at 60% to 70% of heart rate reserve (Meeuwsen et al. 2001). Eight fitness-matched athletes performed similar training in normobaric normoxia. Following the 10-day experimental period, there were no significant increments in hemoglobin or hematocrit for the IHT athletes. Postexperimental $\dot{V}O_2$max was unchanged in both groups; however, maximal power output increased 6% ($p < 0.05$) in the IHT group. In addition, the IHT group demonstrated significant improvements in anaerobic power (5%) and anaerobic capacity (4%) measured via a 30-sec Wingate test. Comparable results for anaerobic power and anaerobic capacity were reported recently by Hendriksen and Meeuwsen (2003) in a study that used a similar IHT protocol. Collectively, these data suggested that 10 days of IHT had no effect on the hematological status and $\dot{V}O_2$max of elite triathletes, but did enhance anaerobic power and capacity. A recent study by Truijens et al. (2003) also evaluated the effect of IHT at a simulated altitude of 2,500 m (8,200 ft). Trained collegiate and masters swimmers completed a 5-week training program consisting of three high-intensity workouts per week in a swimming flume and additional weekly training sessions in a pool. The IHT group completed their high-intensity flume workouts while inspiring a gas mixture with F_IO_2 0.153 (2,500 m/8,200 ft), whereas a control group inspired a gas mixture with F_IO_2 0.209 (sea level) during the flume workouts. Training volume and training intensity were matched between the IHT and control groups. Following the 5-week training program, both groups improved their 100-m and 400-m swim performances by approximately 1% to 2%. However, improvements in swim performance, as well as $\dot{V}O_2$max, were not significantly different for the IHT group compared with the control group (Truijens et al. 2003).

Several investigations have evaluated the efficacy of IHT done at relatively high simulated elevations (3,000-5,500 m/9,840-18,040 ft) (Hahn et al. 1992; Karlsen et al. 2002; Katayama et al. 1999; Rodriguez et al. 1999; Vallier et al. 1996). Using a randomized, blinded crossover design, Karlsen et al. (2002) recently evaluated female and male junior elite cyclists who completed three weeks of IHT, which included three 2-hr IHT interval workouts per week in a nitrogen apartment at a simulated altitude of 3,000 m (9,840 ft). IHT had no effect on the performance of a 30-min cycle ergometer time trial, $\dot{V}O_2$max, maximal power output, and several measures of enhanced erythropoiesis compared with the three weeks of normoxic training. Hahn et al. (1992) evaluated eight female and male elite rowers from the Australian national team. The athletes completed a 19-day training block in which rowing workouts were performed while inspiring supplemental gas with F_IO_2 ~0.15, which simulated an altitude of 3,100 m (10,170 ft). Training consisted of three rowing workouts a day; one of the workouts required the athletes to complete two, 30-minute rowing pieces (3-minute recovery interval) on a rowing ergometer at 70 to 75% HRmax. Eight female and male rowers completed an identical training program while inspiring a supplemental gas with F_IO_2 ~0.21, which simulated sea level conditions. When measured on the first day following the 19-day training period, hemoglobin concentration was 1% lower (NSD) and rowing $\dot{V}O_2$max was nearly identical to pretraining values in the IHT rowers. Rowing performance was evaluated using a 2,500-m rowing ergometer time trial, which was completed on the sixth day following the 19-day training block. The IHT group rowed 1% faster on the posttraining rowing test; however, this improvement was not statistically different compared to either their pretraining rowing test or the posttraining rowing performance of the control group. Thus, this well-controlled study suggested that performing three IHT workouts a day at a simulated altitude equivalent to 3,100 m (10,170 ft) does not alter hemoglobin concentration, maximal oxygen consumption, or rowing performance any differently than similar training at sea level.

Vallier et al. (1996) examined the use of IHT by five female and male elite triathletes who were members of the French national team. The athletes trained three days per week for three weeks in a hypobaric chamber at a simulated altitude of 4,000 m (13,120 ft). Hypobaric training sessions were approximately 60 min in duration and included endurance workouts at 66% of maximal power output ($Power_{max}$) and interval workouts at 85% $Power_{max}$. A non-IHT control group was not included in the experimental design. The athletes were evaluated on the seventh day after completion of the three-week training period. Compared with preexperimental values, no significant differences were observed for hemoglobin (–3%) or sea level $\dot{V}O_2$max (2%). In addition, maximal power output during an incremental exhaustive cycling test was identical following three weeks of IHT compared with the prealtitude test. Similar results were reported by Katayama and colleagues (1999) in a study of trained males who trained on a cycle ergometer for 30 min per day (70% of altitude $\dot{V}O_2$max), five days per week for two weeks in a hypobaric chamber at a simulated altitude of 4,500 m (14,760 ft). A fitness-matched sea level control group followed the identical training regimen at the same relative exercise intensity (70% of sea level $\dot{V}O_2$max) but outside the hypobaric chamber. Following the two-week training period, $\dot{V}O_2$max improved significantly (5% to 7%) in both groups; however, there were no differences between the IHT and control groups. Collectively, the data from these several investigations suggested that IHT conducted at a simulated altitude of 3,000 m to 5,500 m (9,840 ft to 18,040 ft) did not affect hemoglobin levels, $\dot{V}O_2$max, or sea level performance in well-trained athletes.

Rodriguez et al. (1999) evaluated the effect of passive IHE combined with IHT. A group of experienced mountaineers (n = 7) completed a nine-day period during which they were passively exposed in a hypobaric chamber for 3 to 5 hr a day to a simulated altitude that progressed from 4,000 m (13,120 ft) on day 1 to 5,500 m (18,040 ft) on day 9. In addition, each day the subjects performed three to five bouts of cycle ergometer exercise (10-15 min per bout) at a moderate intensity in the hypobaric hypoxic environment. Another group of mountaineers (n = 10) completed the same passive IHE protocol, but did not undertake any intermittent hypoxic exercise training. Following the nine-day experimental period, no significant differences were observed between the groups in any of the hematological or exercise performance parameters. Thus data from the two groups were pooled; the results indicated significant pre- versus postexperimental improvements in reticulocyte percent (54%), RBC (12%), hematocrit (11%), and hemoglobin (18%). During a maximal incremental treadmill test, significant improvements in exercise time (4%) and velocity at the blood lactate threshold were also demonstrated, but $\dot{V}O_2$max was unaffected. Similar results were reported by Casas et al. (2000) in a study that used a similar IHE + IHT design, but over a longer period of time (17 days). These data suggested that intermittent hypoxic exercise training had no additional effect on erythropoiesis and exercise performance compared with passive intermittent hypoxic exposure. Based on the results of Rodriguez et al. (1999) and Casas et al. (2000), it appears that passive IHE alone is sufficient to induce hematological changes and enhance aerobic performance in well-trained mountaineers; however, this conclusion is open to debate since a non-IHE control group was not included in the research design of these studies.

In summary, it is unclear whether IHE or IHT leads to improvements in RBC, hemoglobin, and hematocrit despite increments in serum EPO. There are minimal data to support the claim that IHE or IHT enhances $\dot{V}O_2$max and endurance performance in well-trained athletes. However, preliminary data suggest that anaerobic power and anaerobic capacity may be improved as a result of IHT. Two studies (Green et al. 1999; Melissa et al. 1997) have demonstrated potentially beneficial changes in citrate synthase activity of the vastus lateralis muscle in untrained subjects following eight weeks of IHT. Whether similar changes occur in well-trained or elite athletes is unknown. Table 6.6 summarizes the scientific research on IHE and IHT as it relates to athletic performance.

Table 6.6 Scientific Literature About the Use of Passive IHE or IHT

Author	Altitude (m/ft)	Subjects	Normoxia control	IHE/IHT protocol	Δ EPO[a] (%)	Δ Reticulocytes[a] (%)	Δ RBC[b] (million/mm³)	Δ Hb[b] (%)	Δ V̇O₂max[b] (%)	Postaltitude SL performance[b]
IHE										
Hellemans 1999	5,800-6,400/ 19,025-20,990	Male endurance athletes (n = 10)	No	IHE 5 min "on" + 5 min "off" for 1 hr, two sessions per day for 20 days using a Hypoxicator device	NR	29*	NR	13*	NR	3%* improvement in endurance performance
Rodriguez et al. 1999	4,000-5,500/ 13,120-18,040	Female and male mountaineers (n = 10)	No	IHE for 3-5 hr, one session per day for 9 days using a hypobaric chamber	NR	54*	12*	18*	NSD	4%* improvement in time to exhaustion on a maximal treadmill test
Rodriguez et al. 2000	4,000-5,500/ 13,120-18,040	Male trained athletes (n = 8)	No	IHE for 90 min, three sessions per week for 3 weeks using a hypobaric chamber	NR	108*	7*	13*	NSD	NSD in exercise time to exhaustion on a maximal cycle ergometer test
Glyde-Julian et al. 2003	5,000/16,400	Female and male national level runners (n = 14)	Yes	IHE 5 min "on" + 5 min "off" for 70 min, two sessions per day, 5 days per week for 4 weeks using a Hypoxicator device	NSD	NSD	NSD	NSD	NSD	NSD in 3,000-m TT performance

(continued)

Table 6.6 *(continued)*

Author	Altitude (m/ft)	Subjects	Normoxia control	IHE/IHT protocol	Δ EPO[a] (%)	Δ Reticu- locytes[a] (%)	Δ RBC[b] (million/ mm³)	Δ Hb[b] (%)	Δ V̇O₂max[b] (%)	Postaltitude SL performance[b]
Frey et al. 2000	6,400/20,990	Female and male moderately trained subjects (n = NR)	No	IHE for 75 min, one session per day for 21 days using a Hypoxicator device	38*	NSD	NSD	NSD	NSD	NR
IHT										
Terrados et al. 1988	2,300/7,544	Male elite cyclists (n = 8)	Yes	Cycle IHT for 1-2 hr, one session per day, 4-5 days per week for 3-4 weeks using a hypobaric chamber	NR	NR	NR	8 NSD	NSD	Improvements in total work capacity (33%*) and maximal power output (12%*) on a maximal cycle ergometer test
Meeuwsen et al. 2001 Hendriksen and Meeuwsen 2003	2,500/8,200	Male well-trained triathletes (n = 16)	Yes	Cycle IHT for 2 hr, one session per day for 10 days using a hypobaric chamber	NR	NR	NR	NSD	NSD	Improvements in maximal power output (6%*) on a maximal cycle ergometer test. Improvements in anaerobic power (5%*) and anaerobic capacity (4%*) on a 30-s Wingate test.

Study	Altitude	Subjects	Control	Protocol						Performance outcome
Truijens et al. 2003	2,500/8,200	Female and male trained swimmers (n = 16)	Yes	Swim IHT, three high-intensity interval sessions per week for 5 weeks using hypoxic gas in a swimming flume	NR	NR	NR	NSD	NSD	NSD in 100-m or 400-m swim performance
Karlsen et al. 2002	3,000/9,840	Female and male junior elite cyclists (n = 8)	Yes (crossover design)	Cycle IHT for 2 hr, three high-intensity interval sessions per week for 3 weeks using a nitrogen apartment	NSD	NSD	NSD	NSD	NSD	NSD in 30-min cycle ergometer TT
Hahn et al. 1992	3,100/10,170	Female and male elite rowers (n = 16) Australian NT	Yes	Rowing IHT, three sessions per day for 19 days using hypoxic gas	NR	NR	NR	−1 NSD	NSD	NSD in 2,500-m rowing ergometer TT performance
Vallier et al. 1996	4,000/13,120	Female and male elite triathletes (n = 5) French NT	No	Cycle IHT for 60 min, one session per day, 3 days per week for 3 weeks using a hypobaric chamber	NR	NR	NR	−3 NSD	2 NSD	NSD in maximal power output on a maximal cycle ergometer test
Katayama et al. 1999	4,500/14,760	Male moderately trained athletes (n = 14)	Yes	Cycle IHT for 30 min, one session per day, 5 days per week for 2 weeks using a hypobaric chamber	NR	NR	NR	NR	NSD	NR

(continued)

Table 6.6 (continued)

Author	Altitude (m/ft)	Subjects	Normoxia control	IHE/IHT protocol	Δ EPO[a] (%)	Δ Reticulocytes[a] (%)	Δ RBC[b] (million/mm³)	Δ Hb[b] (%)	Δ $\dot{V}O_2max$[b] (%)	Postaltitude SL performance[b]
Rodriguez et al. 1999	4,000-5,500/ 13,120-18,040	Female and male mountaineers (n = 7)	No	IHE for 3-5 hr, one session per day for 9 days + cycle IHT for 10-15 min, three to five sessions per day over the same 9 days using a hypobaric chamber	NR	54*	12*	18*	NSD	4%* improvement in time to exhaustion on a maximal treadmill test
Casas et al. 2000	4,000-5,500/ 13,120-18,040	Female and male mountaineers (n = 6)	No	IHE for 3-5 hr, one session per day for 17 days + cycle IHT for 10-15 min, three to five sessions per day over the same 17 days using a hypobaric chamber	NR	NR	8*	11*	NSD	NSD in time to exhaustion on a maximal cycle ergometer test. Significant improvement in blood lactate threshold parameters.

Studies are listed by ascending altitude.

Δ = change
[a] Measured within 1-5 days after starting IHE/IHT.
[b] Measured within 1-7 days following IHE/IHT.
* Significant difference vs. pre-IHE/IHT ($p < 0.05$).

EPO = erythropoietin (mU · ml⁻¹); Hb = hemoglobin (g · dL⁻¹); IHE = intermittent hypoxic exposure; IHT = intermittent hypoxic training; NR = not reported/measured; NSD = no statistically significant difference; NT = national team; RBC = red blood cell count (million/mm³); TT = time trial.

Ethical Considerations

The use of some of the altitude training strategies and devices described in this chapter has raised issues regarding the ethical integrity of their utilization for the enhancement of athletic performance. In its most recent position statement (2000) on the use of banned substances, the Medical Commission of the International Olympic Committee (IOC) defined *doping* as "the use of an artifice, whether substance or method, potentially dangerous to athletes' health and/or capable of enhancing their performances, or the presence in the athlete's body of a substance, or the ascertainment of the use of a method on the list annexed to the Olympic Movement Anti-Doping Code" (de Merode 2000). Moreover, *blood doping* is defined by the IOC's Medical Commission as "the administration of blood, RBCs, artificial oxygen carriers and related blood products to an athlete" (de Merode 2000). On the basis of these definitions, some have objected to the use of normobaric hypoxic apartments, supplemental oxygen, altitude tents, and so forth on ethical and legal grounds, claiming that they provide an unfair advantage to those athletes who use them in preparation for athletic competition. Indeed, there was so much public opposition in Norway regarding the use of simulated altitude devices by their Olympic athletes that, in July 2001, the Norwegian Winter Olympic Team vacated the nitrogen house they had constructed in Heber City, Utah, that was being used by the Norwegian athletes in preparation for the 2002 Salt Lake City Olympics. The use of simulated altitude devices by athletes living in the Olympic village during the 2000 Sydney Olympics was prohibited and the same policy will be in place for the 2004 Athens Olympics.

A recent study (Ashenden et al. 2001), however, indicated that the use of a normobaric hypoxic apartment (simulated altitude 2,650-3,000 m/8,690-9,840 ft) by a group of well-trained endurance athletes for 11 to 23 days resulted in a change in percent reticulocytes (30% increase) that was markedly less than that seen in a group of healthy individuals who were given a relatively low dose (150 IU · kg^{-1} per week for 3.5 weeks) of the banned erythropoietic drug, recombinant human EPO (89% increase). It was concluded that simulated altitude facilities should not be considered unethical based strictly on the belief that they produce performance-enhancing changes in RBC mass similar to those produced via illegal substances. In light of this ongoing controversy, the IOC's Medical Commission recently indicated that they will evaluate the safety and ethical issues related to the use of simulated altitude devices. Their findings are expected to be released prior to the 2004 Athens Summer Olympics.

Table 6.7 Contemporary Simulated Altitude Training Methods and Devices

Method	Device	Simulated environment	Scientific principle	Altitude training applications
Nitrogen dilution	Nitrogen apartment	Normobaric hypoxia	Normobaric hypoxic environment simulated via nitrogen dilution ($\downarrow F_IO_2$)	LHTL: Live/sleep in a simulated normobaric hypoxic environment, train in a natural normobaric normoxic environment with minimal travel or inconvenience.
Oxygen filtration	Hypoxico Altitude Tent Colorado Mountain Room	Normobaric hypoxia	Normobaric hypoxic environment simulated via filtration of inspired O_2 molecules ($\downarrow F_IO_2$)	LHTL: Sleep in a simulated normobaric hypoxic environment, train in a natural normobaric normoxic environment with minimal travel or inconvenience.
Hypobaria	Hypobaric chamber Colorado Altitude Training (CAT) Hatch	Hypobaric hypoxia	Hypobaric hypoxic environment simulated via a reduction in barometric pressure ($\downarrow PO_2$)	LHTL: Sleep in a simulated hypobaric hypoxic environment, train in a natural normobaric normoxic environment with minimal travel or inconvenience.
Supplemental oxygen	Hyperoxic medical grade gas	Normobaric normoxia	Normobaric normoxic environment simulated via use of supplemental O_2 ($\uparrow F_IO_2$) at altitude	LH + TLO$_2$: Live/sleep in a natural hypobaric hypoxic environment, train in a simulated normobaric normoxic environment with minimal travel or inconvenience.
Intermittent hypoxic exposure (IHE)	Hypobaric chamber	Hypobaric hypoxia	Hypobaric hypoxic environment simulated via a reduction in barometric pressure ($\downarrow PO_2$)	LLTH: Live in a natural normobaric normoxic environment, passively exposed intermittently to a hypobaric hypoxic or normobaric hypoxic environment.
	Hypoxicator/Go2Altitude unit	Normobaric hypoxia	Normobaric hypoxic environment simulated via filtration of inspired O_2 molecules ($\downarrow F_IO_2$)	
	Hypoxic medical grade gas	Normobaric hypoxia	Normobaric hypoxic environment simulated via a reduction of inspired O_2 concentration ($\downarrow F_IO_2$)	
Intermittent hypoxic exposure (IHT)	Hypobaric chamber	Hypobaric hypoxia	Hypobaric hypoxic environment simulated via a reduction in barometric pressure ($\downarrow PO_2$)	LLTH: Live in a natural normobaric normoxic environment, train intermittently in a hypobaric hypoxic or normobaric hypoxic environment.
	Hypoxicator/Go2Altitude unit	Normobaric hypoxia	Normobaric hypoxic environment simulated via filtration of inspired O_2 molecules ($\downarrow F_IO_2$)	
	Hypoxic medical grade gas	Normobaric hypoxia	Normobaric hypoxic environment simulated via a reduction of inspired O_2 concentration ($\downarrow F_IO_2$)	

F_IO_2 = fraction of inspired oxygen, LHTL = live high-train low, LH + TLO$_2$ = live high-train low using supplemental oxygen, LLTH = live low-train high, PO_2 = partial pressure of oxygen

Summary

Development in the early 1990s of LHTL has led to the use of several new altitude training strategies and devices by elite athletes. These include normobaric hypoxic apartments, supplemental oxygen, and hypoxic sleeping units. Several investigations have demonstrated that chronic use of a normobaric hypoxic apartment (12 to 18 hr per day for 10 to 25 days at 2,000 m to 3,000 m/6,560 ft to 9,840 ft) stimulates the release of serum EPO and significantly increases reticulocyte count. In turn, these erythropoietic changes have been associated with improvements in postaltitude $\dot{V}O_2$max and endurance performance. However, other studies have failed to demonstrate any significant erythropoietic effect as a result of normobaric hypoxic exposure. These discrepancies between studies may be the result of differences in assessment methods, hypoxic stimulus, or the athlete's training status. Although limited, published data regarding the efficacy of supplemental oxygen training suggest that high-intensity workouts at moderate altitude and endurance performance at sea level may be enhanced via hyperoxic training utilized over a period of several weeks. At present, no scientific studies have been published that document the effect of hypoxic sleeping devices on erythropoiesis, $\dot{V}O_2$max, or performance in athletes.

Intermittent hypoxic exposure/training allows athletes to "live low-train high." It is unclear whether IHE or IHT leads to improvements in RBC, hematocrit, and hemoglobin. In addition, there are minimal data to support the claim that IHE or IHT enhances $\dot{V}O_2$max and endurance performance. Preliminary data, however, suggest that anaerobic power and anaerobic capacity may be improved as a result of IHT.

Table 6.7 on page 218 summarizes the characteristics of current simulated altitude training methods and devices described in this chapter.

References

Ashenden, M.J., C.J. Gore, G.P. Dobson, T.T. Boston, R. Parisotto, K.R. Emslie, G.J. Trout, and A.G. Hahn. 2000. Simulated moderate altitude elevates serum erythropoietin but does not increase reticulocyte production in well-trained runners. *European Journal of Applied Physiology* 81: 428-435.

Ashenden, M.J., C.J. Gore, G.P. Dobson, and A.G. Hahn. 1999a. "Live high, train low" does not change the total haemoglobin mass of male endurance athletes sleeping at a simulated altitude of 3000 m for 23 nights. *European Journal of Applied Physiology* 80: 479-484.

Ashenden, M.J., C.J. Gore, D.T. Martin, G.P. Dobson, and A.G. Hahn. 1999b. Effects of a 12-day "live high, train low" camp on reticulocyte production and haemoglobin mass in elite female road cyclists. *European Journal of Applied Physiology* 80: 472-478.

Ashenden, M.J., A.G. Hahn, D.T. Martin, P. Logan, R. Parisotto, and C.J. Gore. 2001. A comparison of the physiological response to simulated altitude exposure and r-HuEpo administration. *Journal of Sports Sciences* 19: 831-837.

Bartsch, P. 1992. Treatment of high altitude diseases without drugs. *International Journal of Sports Medicine* 13 (Suppl. 1): S71-S74.

Casas, M., H. Casas, T. Pages, R. Rama, A. Ricart, J.L. Ventura, J. Ibanez, F.A. Rodriguez, and G. Viscor. 2000. Intermittent hypobaric hypoxia induces altitude acclimation and improves lactate threshold. *Aviation, Space and Environmental Medicine* 71: 125-130.

Chick, T.W., D.M. Stark, and G.H. Murata. 1993. Hyperoxic training increases work capacity after maximal training at moderate altitude. *Chest* 104: 1759-1762.

de Merode, A. 2000. International Olympic Committee Medical Commission prohibited classes of substances and methods of doping, 2000 Feb 15 (letter). In Bowers, L.D., S.M. Bailey, and J. Podraza, *United States Anti-Doping Agency guide to prohibited classes of substances and prohibited methods of doping*, p. 3. Colorado Springs, CO: US Anti-Doping Agency.

Eckardt, K.-U., U. Boutellier, A. Kurtz, M. Schopen, E.A. Koller, and C. Bauer. 1989. Rate of erythropoietin formation in humans in response to acute hypobaric hypoxia. *Journal of Applied Physiology* 66: 1785-1788.

Frey, W.O., R. Zenhausern, P.C. Colombani, and J. Fehr. 2000. Influence of intermittent exposure to normobaric hypoxia on hematological indexes and exercise performance. *Medicine and Science in Sports and Exercise* 32 (Suppl. 5): S65.

Gamow, R.I., G.D. Geer, J.F. Kasic, and H.M. Smith. 1990. Methods of gas-balance control to be used with a portable hyperbaric chamber in the treatment of high altitude illness. *Wilderness Medicine* 1: 165-180.

Glyde-Julian, C.G., R.L. Wilber, J.T. Daniels, M. Fredericson, J. Stray-Gundersen, A.G. Hahn, R.T. Parisotto, C.J. Gore, and B.D. Levine. 2003. Intermittent normobaric hypoxia does not alter performance or erythropoietic markers in elite distance runners. *Journal of Applied Physiology* (in review).

Gore, C.J., A.G. Hahn, R.J. Aughey, D.T. Martin, M.J. Ashenden, S.A. Clark, A.P. Garnham, A.D. Roberts, G.J. Slater, and M.J. McKenna. 2001. Live high:train low increases muscle buffer capacity and submaximal cycling efficiency. *Acta Physiologica Scandinavica* 173: 275-286.

Graham, T.E., P.K. Petersen, and B. Saltin. 1987. Muscle and blood ammonia and lactate responses to prolonged exercise with hyperoxia. *Journal of Applied Physiology* 63: 1457-1462.

Green, H., J. MacDougall, M.A. Tarnapolsky, and N.L. Melissa. 1999. Downregulation of Na^+-K^+-ATPase pumps in skeletal muscle with training in normobaric hypoxia. *Journal of Applied Physiology* 86: 1745-1748.

Hahn, A.G., R.D. Telford, D.M. Tumilty, M.E. McBride, D.P. Campbell, and J.C. Kovacic. 1992. Effect of supplementary hypoxic training on physiological characteristics and ergometer performance of elite rowers. *Excel* 8: 127-138.

Hamalainen, I.T., A.T. Nummela, and H.K. Rusko. 2000. Training in hyperoxia improves 3000-m running performance in national level athletes. *Medicine and Science in Sports and Exercise* 32 (Suppl. 5): S47.

Hellemans, J. 1999. Intermittent hypoxic training: A pilot study. *Proceedings of the Second Annual International Altitude Training Symposium.* Flagstaff, AZ.

Hendriksen, I.J.M., and T. Meeuwsen. 2003. The effect of intermittent training in hypobaric hypoxia on sea-level exercise: A cross-over study in humans. *European Journal of Applied Physiology* 88: 396-403.

Howley, E.T., R.H. Cox, H.G. Welch, and R.P. Adams. 1983. Effect of hyperoxia on metabolic and catecholamine responses to prolonged exercise. *Journal of Applied Physiology* 54: 59-63.

Hughson, R.L., and J.M. Kowalchuk. 1995. Kinetics of oxygen uptake for submaximal exercise in hyperoxia, normoxia, and hypoxia. *Canadian Journal of Applied Physiology* 20: 198-210.

Ingham, E.A., P.D. Pfitzinger, J. Hellemans, C. Bailey, J.S. Fleming, and W.G. Hopkins. 2001. Running performance following intermittent altitude exposure simulated with nitrogen tents. *Medicine and Science in Sports and Exercise* 33 (Suppl. 5): S2.

Karlsen, T., O. Madsen, S. Rolf, and J. Stray-Gundersen. 2002. Effects of 3 weeks hypoxic interval training on sea level cycling performance and hematological parameters. *Medicine and Science in Sports and Exercise* 34 (Suppl. 5): S224.

Katayama, K., Y. Sato, Y. Morotome, N. Shima, K. Ishida, S. Mori, and M. Miyamura. 1999. Ventilatory chemosensitive adaptations to intermittent hypoxic exposure with endurance training and detraining. *Journal of Applied Physiology* 86: 1805-1811.

Knaupp, W., S. Khilnani, J. Sherwood, S. Scharf, and H. Steinberg. 1992. Erythropoietin response to acute normobaric hypoxia in humans. *Journal of Applied Physiology* 73: 837-840.

Laitinen, H., K. Alopaeus, R. Heikkinen, H. Hietanen, L. Mikkelsson, H. Tikkanen, and H.K. Rusko. 1995. Acclimatization to living in normobaric hypoxia and training at sea level in runners. *Medicine and Science in Sports and Exercise* 27 (Suppl. 5): S109.

Linossier, M.-T., D. Dormois, L. Arsac, C. Denis, J.-P. Gay, A. Geyssant, and J.-R. Lacour. 2000. Effect of hyperoxia on aerobic and anaerobic performances and muscle metabolism during maximal cycling exercise. *Acta Physiologica Scandinavica* 168: 403-411.

Martin, D.T., A.G. Hahn, H. Lee, A.D. Roberts, J. Victor, and C.J. Gore. 2002. Effects of a 12-day "live high, train low" cycling camp on 4-min and 30-min performance. *Medicine and Science in Sports and Exercise* 34 (Suppl. 5): S274.

Mattila, V., and H. Rusko. 1996. Effect of living high and training low on sea level performance in cyclists. *Medicine and Science in Sports and Exercise* 28 (Suppl. 5): S157.

Meeuwsen, T., I.J.M. Hendriksen, and M. Holewijn. 2001. Training-induced increases in sea-level performance are enhanced by acute intermittent hypobaric hypoxia. *European Journal of Applied Physiology* 84: 283-290.

Melissa, L.M., J.D. MacDougall, M.A. Tarnapolsky, N. Cipriano, and H.J. Green. 1997. Skeletal muscle adaptations to training under normobaric hypoxic versus normoxic conditions. *Medicine and Science in Sports and Exercise* 29: 238-243.

Mizuno, M., C. Juel, T. Bro-Rasmussen, E. Mygind, B. Schibye, B. Rasmussin, and B. Saltin. 1990. Limb skeletal muscle adaptations in athletes after training at altitude. *Journal of Applied Physiology* 68: 496-502.

Morris, D.M., J.T. Kearney, and E.R. Burke. 2000. The effects of breathing supplemental oxygen during altitude training on cycling performance. *Journal of Science and Medicine in Sport* 3: 165-175.

Nummela, A., M. Alberts, R.P. Rijntjes, P. Luhtanen, and H. Rusko. 1996a. Reliability and validity of the maximal anaerobic running test. *International Journal of Sports Medicine* 17 (Suppl. 2): S97-S102.

Nummela, A., I.T. Hamalainen, and H.K. Rusko. 2000. Effect of hyperoxia on S_aO_2, blood pH and heart rate recovery during intermittent exercise. *Medicine and Science in Sports and Exercise* 32 (Suppl. 5): S358.

Nummela, A., M. Mero, J. Stray-Gundersen, and H. Rusko. 1996b. Important determinants of anaerobic running performance in male athletes and non-athletes. *International Journal of Sports Medicine* 17 (Suppl. 2): S91-S96.

Nummela, A., and H. Rusko. 2000. Acclimatization to altitude and normoxic training improve 400-m running performance at sea level. *Journal of Sports Sciences* 18: 411-419.

Peltonen, J.E., A.P. Leppavuori, K.-P. Kyro, P. Makela, and H.K. Rusko. 1999. Arterial haemoglobin oxygen saturation is affected by F_IO_2 at submaximal running velocities in elite athletes. *Scandinavian Journal of Medicine and Science in Sports* 9: 265-271.

Peltonen, J.E., J. Rantamaki, S.P.T. Niittymaki, K. Sweins, J.T. Viitasalo, and H.K. Rusko. 1995. Effects of oxygen fraction in inspired air on rowing performance. *Medicine and Science in Sports and Exercise* 27: 573-579.

Piehl-Aulin, K., J. Svedenhag, L. Wide, B. Berglund, and B. Saltin. 1998. Short-term intermittent normobaric hypoxia—haematological, physiological and mental effects. *Scandinavian Journal of Medicine and Science in Sports* 8: 132-137.

Poulsen, T.D., T. Klausen, J.-P. Richalet, I.-L. Kanstrup, N. Fogh-Andersen, and N.V. Olsen. 1998. Plasma volume in acute hypoxia: comparison of a carbon monoxide rebreathing method and dye dilution with Evan's blue. *European Journal of Applied Physiology* 77: 457-461.

Powell, F.L., and N. Garcia. 2000. Physiological effects of intermittent hypoxia. *High Altitude Medicine and Biology* 1: 125-136.

Powers, S.K., J. Lawler, J.A. Dempsey, S. Dodd, and G. Landry. 1989. Effects of incomplete pulmonary gas exchange on $\dot{V}O_2$max. *Journal of Applied Physiology* 66: 2491-2495.

Roberts, A.D., S.A. Clark, N.E. Townsend, M.E. Anderson, C.J. Gore, and A.G. Hahn. 2003. Changes in performance, maximal oxygen uptake and maximal accumulated oxygen deficit after 5, 10 and 15 days of live high:train low altitude exposure. *European Journal of Applied Physiology* 88: 390-395.

Rodriguez, F.A., T. Cabanes, X. Iglesias, M. Huertas, H. Casas, and J.L. Ventura. 2002a. Intermittent hypobaric hypoxia enhances cycling performance in world-class track cyclists. In Koskolou, M., N. Geladas, and V. Klissouras (eds.), *Proceedings of the 7th Annual Congress of the European College of Sport Science, vol. 1*, p. 83. Athens: European College of Sport Sciences, University of Athens.

Rodriguez, F.A., H. Casas, M. Casas, T. Pages, R. Rama, A. Ricart, J.L. Ventura, J. Ibanez, and G. Viscor. 1999. Intermittent hypobaric hypoxia stimulates erythropoiesis and improves aerobic capacity. *Medicine and Science in Sports and Exercise* 31: 264-268.

Rodriguez, F.A., J. Murio, H. Casas, G. Viscor, and J.L. Ventura. 2002b. Intermittent hypobaric hypoxia enhances swimming performance and maximal aerobic power in trained swimmers. In Koskolou, M., N. Geladas, and V. Klissouras (eds.), *Proceedings of the 9th World Symposium on Biomechanics and Medicine in Swimming*, p. 152. Saint-Etienne, France.

Rodriguez, F.A., J.L. Ventura, M. Casas, H. Casas, T. Pages, R. Rama, A. Ricart, L. Palacios, and G. Viscor. 2000. Erythropoietin acute reaction and hematological adaptations to short, intermittent hypobaric hypoxia. *European Journal of Applied Physiology* 82: 170-177.

Rusko, H.K., A. Leppavuori, P. Makela, and J. Leppaluoto. 1995. Living high, training low: A new approach to altitude training at sea level in athletes. *Medicine and Science in Sports and Exercise* 27 (Suppl. 5): S6.

Rusko, H.K., H. Tikkanen, L. Paavolainen, I. Hamalainen, K. Kalliokoski, and A. Puranen. 1999. Effect of living in hypoxia and training in normoxia on sea level $\dot{V}O_2$max and red cell mass. *Medicine and Science in Sports and Exercise* 31 (Suppl. 5): S86.

Schmidt, W. 2002. Effects of intermittent exposure to high altitude on blood volume and erythropoietic activity. *High Altitude Medicine and Biology* 3: 167-176.

Shannon, M.P., R.L. Wilber, and J.T. Kearney. 2001. Normobaric-hypoxia: Performance characteristics of simulated altitude tents. *Medicine and Science in Sports and Exercise* 33 (Suppl. 5): S60.

Terrados, N., J. Melichna, C. Sylven, E. Jansson, and L. Kaijser. 1988. Effects of training at simulated altitude on performance and muscle metabolic capacity in competitive road cyclists. *European Journal of Applied Physiology* 57: 203-209.

Townsend, N.E., C.J. Gore, A.G. Hahn, M.J. McKenna, R.J. Aughey, S.A. Clark, T. Kinsman, J.A. Hawley, and C.-M. Chow. 2002. Living high-training low increases hypoxic ventilatory response in well-trained endurance athletes. *Journal of Applied Physiology* 93: 1498-1505.

Truijens, M.J., H.M. Toussaint, J. Dow, and B.D. Levine. 2003. Effect of high-intensity hypoxic training on sea-level swim performances. *Journal of Applied Physiology* 94: 733-743.

Vallier, J.M., P. Chateau, and C.Y. Guezennec. 1996. Effects of physical training in a hypobaric chamber on the physical performance of competitive triathletes. *European Journal of Applied Physiology* 73: 471-478.

Welch, H.G. 1987. Effects of hypoxia and hyperoxia on human performance. *Exercise and Sport Sciences Reviews* 15: 191-220.

Wilber, R.L. 2001. Current trends in altitude training. *Sports Medicine* 31: 249-265.

Wilber, R.L., P.L. Holm, D.M. Morris, G.M. Dallam, and S.D. Callan. 2003a. Effect of F_IO_2 on oxidative stress during high-intensity interval training at moderate altitude. *Medicine and Science in Sports and Exercise* 35 (Suppl. 5): S116.

Wilber, R.L., P.L. Holm, D.M. Morris, G.M. Dallam, and S.D. Callan. 2003b. Effect of F_IO_2 on physiological responses and cycling performance at moderate altitude. *Medicine and Science in Sports and Exercise* 35: 1153-1159.

Wilber, R.L., M.P. Shannon, J.T. Kearney, P.L. Holm, and M.R. Hill. 2002. Operational characteristics of a normobaric hypoxic system. Proceedings of the Sixth International Olympic Committee World Congress on Sport Sciences. In *Medicine and Science in Sports and Exercise* 34 (Suppl. 5): 92.

Chapter 7
RECOMMENDATIONS AND GUIDELINES

Based on scientific data as well as experiential evidence from athletes and coaches, the following recommendations and guidelines are offered for effective altitude training in preparation for sea level performance. It is important to reemphasize that each athlete responds differently to living and training at altitude. Therefore, these general recommendations and guidelines should serve as a starting point for effectively designing an individualized altitude training program for each athlete.

The "Million-Dollar" Questions

1. What is the optimal altitude at which to live?

    It appears that the optimal elevation at which to live is somewhere between 2,100 m (6,890 ft) and 2,500 m (8,200 ft). In general, locations below 2,100 m may not be high enough to stimulate an increase in red blood cell production, whereas living above 2,500 m may lead to training and recovery problems (Ri-Li et al. 2002; Witkowski et al. 2001). Ideally, athletes would "live/sleep high" between 2,100 m and 2,500 m and would "train low." Use of simulated altitude devices (nitrogen-diluted apartments, altitude tents) or supplemental oxygen makes "live high-train low" (LHTL) a viable option for athletes as opposed to traditional "live high-train high" altitude training.

2. How long does the athlete need to stay at altitude to gain benefits?

    It is recommended that athletes spend a minimum of four weeks at altitude if they expect to derive some of the hematological and muscle buffering benefits associated with altitude training (Gore et al. 2001; Levine and Stray-Gundersen 1997; Stray-Gundersen et al. 2001). Again, the best scenario would be for the athlete to live between 2,100 m and 2,500 m (either naturally or via simulated altitude) and train at a relatively low elevation. If the athlete is using a nitrogen apartment or an altitude tent, exposure to the simulated altitude environment should be for a minimum of 8 to 10 hr per day. Periodic use of supplemental oxygen for high-intensity workouts is recommended (Wilber et al. 2003a,

2003b). Currently, it is not clear whether intermittent hypoxic exposure (IHE) or intermittent hypoxic training (IHT) can serve as an effective alternative to ≥4 consecutive weeks of altitude exposure.

3. How long does the "altitude effect" last after the athlete returns to sea level?

As we would expect, there appears to be tremendous individual variability in the duration of the "altitude effect" after return to sea level. As previously described in this book, Gore et al. (1998) reported that postaltitude cycling performance (4,000-m time trial) was very individual among elite track cyclists when evaluated on days 4, 9, and 21 postaltitude, suggesting that optimal performance for each athlete may occur at a different time point in the weeks following altitude training. In another study (Levine and Stray-Gundersen 1997), 5-km running performance in trained distance runners who completed four weeks of LHTL was similar on days 7, 14, and 21 postaltitude compared with day 3 postaltitude, suggesting that the beneficial effects of LHTL on running performance may last for up to three weeks postaltitude. It is recommended that athletes and coaches perform several "dry runs" well in advance of major competitions to determine the optimal timing of return to sea level following an altitude training phase.

Training Modifications

As discussed in previous chapters, exposure to altitude leads to decrements in the partial pressure of inspired oxygen (P_IO_2), the partial pressure of oxygen in arterial blood (P_aO_2), arterial oxyhemoglobin saturation (S_aO_2), and maximal oxygen consumption ($\dot{V}O_2max$). These altitude-induced physiological responses have a significant impact on an athlete's ability to train and compete at altitude, especially in the first two weeks of altitude exposure. Consequently, it is imperative to make modifications to the athlete's training volume and training intensity to ensure that the athlete does not overtrain. Figure 7.1 shows the recommended modifications to training *volume* at moderate altitude (1,800-3,050 m/5,900-10,000 ft) during the initial five weeks of training. As a general rule, training volume should be reduced by approximately 10% to 20% initially. Figure 7.2a shows the recommended modifications to training *intensity* during interval workouts at moderate altitude during the initial five weeks of training. As a general rule, "sea level" interval workout training intensity should be reduced by approximately 5% to 7% initially. Finally, figure 7.2b shows the recommended modifications

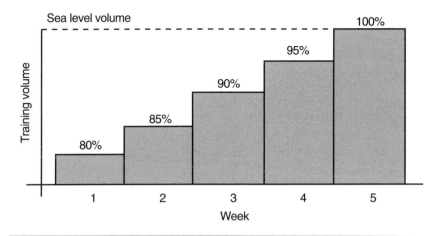

Figure 7.1 Modifications to training volume at moderate altitude (1,800-3,050 m/ 5,900-10,000 ft) during the initial five weeks of training. As a general rule, "sea level" training volume should be reduced by approximately 10% to 20% initially.

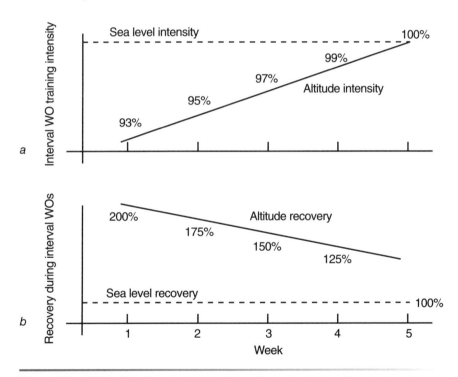

Figure 7.2 (a) Modifications to training intensity during interval workouts at moderate altitude (1,800-3,050 m/5,900-10,000 ft) during the initial five weeks of training. As a general rule, "sea level" interval workout training intensity should be reduced by approximately 5% to 7% initially. (b) Modifications to recovery during interval workouts at moderate altitude (1,800-3,050 m/5,900-10,000 ft) during the initial five weeks of training. As a general rule, "sea level" recovery during interval workouts should be doubled initially.

to *recovery* during interval workouts at moderate altitude over the initial five weeks of training. As a general rule, "sea level" recovery during interval workouts should be doubled initially. Because there is great individual variability among athletes' responses at altitude, some athletes may be able to replicate sea level training volume, intensity, or both within a few weeks at altitude without any negative effects. Other athletes, however, may never be able to replicate sea level workouts at altitude and may need more time to adjust than recommended in the timetables shown in figures 7.1 and 7.2.

Away From Training

In addition to modifications related to training volume and training intensity, athletes and coaches should follow "non-training" guidelines in order to ensure a productive altitude training experience. Most of these steps are either a continuation or a modification of what the athlete would normally do at sea level. However, some of these recommendations are unique to altitude. Table 7.1 summarizes the important non-training issues that athletes and coaches should address prior to and during altitude training.

Altitude Training Sites

Several altitude training sites around the world offer relatively comfortable living and training conditions. Table 7.2 on page 230 provides a list of most of the primary international altitude training sites.

Altitude Training Resources

The following are resources that provide information specific to altitude and products related to altitude training.

Hornbein, T.F., and R.B. Schoene, eds. 2001. *High altitude: An exploration of human adaptation.* New York-Basel: Dekker.

Reeves, J.T., and R.F. Grover. 2001. *Attitudes on altitude: Pioneers of medical research in Colorado's high mountains.* Boulder, CO: University of Colorado Press.

West, J.B., ed. *High altitude medicine and biology.* A professional journal dedicated exclusively to the latest advances in high altitude life sciences. Larchmont, NY: Mary Ann Liebert, Inc. www.liebertpub.com

www.hypoxico.com. Provides information on the Hypoxico Altitude Tent

www.altitudetraining.com. Provides information on the Colorado Altitude Room

www.go2altitude.com. Provides information on Hypoxicator/Go2Altitude device used in intermittent hypoxic exposure (IHE) and intermittent hypoxic training (IHT)

Table 7.1 Recommendations for "Non-Training" Altitude Issues

	Effect of altitude	Recommendation
Heart rate	May be higher versus sea level at rest and during submaximal exercise. May be similar or lower versus sea level during maximal exercise.	Adjustments to heart rate–based "training zones" may be necessary in order to avoid undertraining or overtraining.
Hydration level	Increased respiratory and urinary water loss at altitude. Potential dehydration.	A concerted effort should be made to increase fluid intake by 4 to 5 L per day. Restrict intake of caffeinated beverages.
Carbohydrate metabolism	Increased carbohydrate utilization at altitude. Potential glycogen depletion.	A concerted effort should be made to replace carbohydrate before, during, and after workouts via carbohydrate replacement drinks and solid carbohydrate.
Iron	Decrease in iron stores (ferritin) upon altitude exposure. Potential iron deficiency. Potential non-erythropoietic response.	Have iron status evaluated via a blood test several weeks prior to altitude. Normal iron status: serum ferritin > 20 ng · ml^{-1} in females, > 30 ng · ml^{-1} in males. If iron status is abnormally low prior to altitude, consult your physician regarding iron supplementation.
Immune system	Potential for immunosuppression and increased chance of illness.	Ensure adequate rest and recovery. Ensure adequate postexercise carbohydrate intake to modulate training-induced stress hormone response. Ensure adequate dietary intake of the vitamins (folate, B_6, B_{12}) and minerals (zinc, selenium, copper) that affect the immune system.
Oxidative stress	Increased versus sea level.	Antioxidant supplementation (vitamin A, C, or E).
Ultraviolet radiation	Increased versus sea level.	Ultraviolet sunscreen and sunglasses; antioxidant (vitamin E).
Body composition	Potential decrease in muscle mass at altitude, but not very common at elevations that athletes typically train at (1,800-3,050 m / 5,900-10,000 ft).	Adequate nutrition with possible increase in caloric intake; regular monitoring of body composition.
Sleep/Recovery	May be disturbed and irregular at altitude, especially the first few nights.	Afternoon naps may help until the nighttime sleep disturbances subside. Make sleeping environment as comfortable and as similar to home as possible.
Acute mountain sickness	Potential for symptoms of headache and nausea, but not very common at elevations that athletes typically train at (1,800-3,050 m / 5,900-10,000 ft).	Prescription acetazolamide or aspirin in combination with adequate rest and recovery. Elimination of training or significant modification to training until recovery.

Table 7.2 International Altitude Training Sites

Altitude training site	Country	Elevation (m/ft)
Thredbo Alpine Training Center	Australia	1,365/4,478
Crans Montana	Switzerland	1,500/4,920
Albuquerque, New Mexico	USA	1,525/5,000
Fort Collins, Colorado	USA	1,525/5,000
Davos	Switzerland	1,560/5,117
Issyk-Kull	Kirgizistan	1,600/5,248
Denver, Colorado	USA	1,610/5,280
Medeo	Kazakstan	1,691/5,546
Tamga	Kirgizistan	1,700/5,576
Boulder, Colorado	USA	1,770/5,800
Ifrane	Morocco	1,820/5,970
St. Moritz	Switzerland	1,820/5,970
Nairobi	Kenya	1,840/6,035
Font Romeu Odeillo	France	1,850/6,069
Colorado Springs, Colorado	USA	1,860/6,100
Kunming	China	1,895/6,216
Pontresina	Switzerland	1,900/6,232
Zetersfeld/Linz	Austria	1,950/6,396
Piatra Arsa	Romania	1,950/6,396
Tzahkadzor	Armenia	1,970/6,462
Belmeken	Bulgaria	2,000/6,560
Kesenoy-Am	Russia	2,000/6,560
Sestriere	Italy	2,035/6,675
Flagstaff, Arizona	USA	2,134/7,000
Los Alamos, New Mexico	USA	2,208/7,240
Quito	Ecuador	2,218/7,275
Alamosa, Colorado	USA	2,300/7,544
Mexico City	Mexico	2,300/7,544
Sierra Nevada/Granada	Spain	2,320/7,610
Addis Ababa	Ethiopia	2,400/7,872
Park City, Utah	USA	2,440/8,000
Mammoth Lake, California	USA	2,440/8,000
Bogota	Colombia	2,500/8,200
Toluca	Mexico	2,700/8,856
La Paz	Bolivia	3,100/10,168

Sites are listed by ascending altitude.

Note: Additional Kenyan altitude training sites are listed in table 5.7 (p. 172).

References

Gore, C.J., N.P. Craig, A.G. Hahn, A.J. Rice, P.C. Bourdon, S.R. Lawrence, C.B.V. Walsh, P.G. Barnes, R. Parisotto, D.T. Martin, and D.B. Pyne. 1998. Altitude training at 2690m does not increase total haemoglobin mass or sea level $\dot{V}O_2$max in world champion track cyclists. *Journal of Science and Medicine in Sport* 1: 156-170.

Gore, C.J., A.G. Hahn, R.J. Aughey, D.T. Martin, M.J. Ashenden, S.A. Clark, A.P. Garnham, A.D. Roberts, G.J. Slater, and M.J. McKenna. 2001. Live high:train low increases muscle buffer capacity and submaximal cycling efficiency. *Acta Physiologica Scandinavica* 173: 275-286.

Levine, B.D., and J. Stray-Gundersen. 1997. "Living high-training low": Effect of moderate-altitude acclimatization with low-altitude training on performance. *Journal of Applied Physiology* 83: 102-112.

Ri-Li, G., S. Witkowski, Y. Zhang, C. Alfrey, M. Sivieri, T. Karlsen, G.K. Resaland, M. Harber, J. Stray-Gundersen, and B.D. Levine. 2002. Determinants of erythropoietin release in response to short-term hypobaric hypoxia. *Journal of Applied Physiology* 92: 2361-2367.

Stray-Gundersen, J., R.F. Chapman, and B.D. Levine. 2001. "Living high-training low" altitude training improves sea level performance in male and female elite runners. *Journal of Applied Physiology* 91: 1113-1120.

Wilber, R.L., P.L. Holm, D.M. Morris, G.M. Dallam, and S.D. Callan. 2003a. Effect of F_IO_2 on oxidative stress during high-intensity interval training at moderate altitude. *Medicine and Science in Sports and Exercise* 35 (Suppl. 5): S116.

Wilber, R.L., P.L. Holm, D.M. Morris, G.M. Dallam, and S.D. Callan. 2003b. Effect of F_IO_2 on physiological responses and cycling performance at moderate altitude. *Medicine and Science in Sports and Exercise* 35: 1153-1159.

Witkowski, S., T. Karlsen, G. Resaland, M. Sivieri, R. Yates, M. Harber, R.L. Ge, J. Stray-Gundersen, and B.D. Levine. 2001. Optimal altitude for "living high—training low." *Medicine and Science in Sports and Exercise* 33 (Suppl. 5): S292.

INDEX

Note: Page numbers followed by *f* or *t* refer to the figure or table on that page, respectively.

A

acclimatization. *See also* athletic performance
  and aerobic performance 125–126*t*
  at Arturo Barrios training camp 146
  and blood lactate levels 53–54
  at Dr. Joe Vigil training camp 141–142
  and event duration 127*f*
  and heart rate 50
  Mark Plaatjes' program 150
  and responders/nonresponders to LHTL 110*f*
  of U.S. athletes for 2002 Olympics 130, 131*t*
  of ventilatory system 52
acute mountain sickness 66, 206, 229*t*
Adams, W.C., et al. (1975) 35*t*, 45*t*, 88, 94*t*, 125*t*
adenosine triphosphatase 63
  and LHTH strategy 102
adenosine triphosphate (ATP) 193
aerobic performance 39
  and acclimatization 125–126*t*
  elite athletes at elevations >2,000 m 33, 36
  elite athletes at elevations >3,000 m 37
  elite athletes at moderate altitude 33
  mathematical model of altitude effect 37–38
  maximum cycle ergometry 33
aerodynamics 39
alactate speed training 161
altitude tent 176, 179
altitude training
  duration of effect 226
  factors affecting study results 83–85
  guidelines and recommendations 181*t*
  length of stay 225–226
  LHTL strategy xxiii
  modifications to training 226–228
  optimal altitude 225
  and performance xv–xvi
  physiological benefits 9–10, 16*f*
  rationale xv–xvi
anaerobic performance 46–50
  hypoxic apartment study 187
  and intermittent hypoxic exposure during exercise 210
  and LHTH strategy 104–105
anecdotal evidence. *See* training programs
Arsac, L.M. (2002) 48
arterial oxygen content (C_aO_2) 6
arterial-venous O_2 difference 4
artificial altitude environment. *See* simulation of altitude change
Asano, K., et al. (1986) 91*t*, 100*t*
Ashenden, M.J., et al. (1999a) 190, 196*t*
Ashenden, M.J., et al. (1999b) 189, 195*t*
Ashenden, M.J., et al. (2000) 189, 190, 195*t*
athletic performance
  early research 119–122
  event duration and acclimatization 124
  and LHTH strategy 86–89
  men's 10,000 m at Sydney Olympics xvi
  recent research 122–123
  track and field at Mexico City xx
  world records at Mexico City Olympics xvii
ATP (adenosine triphosphate) 193

B

Bahrke, M.S., and B. Shukitt-Hale (1993) 67
Bailey, D.M., and B. Davies (1997) 83
Bailey, D.M., et al. (1998) 24, 33, 34*t*, 44*t*, 59, 93*t*, 98*t*
Balsom, P.D., et al. (1994) 49*t*
barometric pressure
  altitude-related changes 8*f*
  in Colorado Springs 197–198
  in nitrogen apartment 183
Barrios, Arturo 144–149
Beamon, Bob xvi
Beidleman, B.A., et al. (2002) 65
Berglund, B., et al. (1992) 99*t*
beta-hydroxyacyl-coA dehydrogenase (HADH) 102, 169
Billat, V.L., et al. (2003) 30, 173
Birkeland, K.I., et al. (2000) 11, 12
birth control pill 65
blood doping 9, 13, 16
blood lactate 53–54
  and increased epinephrine 61
blood pH 187
blood volume 51, 171
body composition 63–64, 229*t*
Boning, D. (1997) 83
Brosnan, M.J., et al. (2000) 27*t*, 33, 44*t*, 49*t*
buffering capacity 14, 17, 54, 187
Burtscher, M., et al. (1996) 90*t*
Buskirk, E.R., et al. (1967) 35*t*, 36, 45*t*, 84, 95*t*, 121, 126*t*

C

caloric intake 63–64
Capelli, C., and P.E. di Prampero (1995) 39
capillarity and LHTH strategy 97, 101
capillary length density 133
carbohydrate metabolism 55–56
  and non-training issues 229*t*
  of women at altitude 64

233

carbon dioxide 54
carbon monoxide rebreathing technique 96, 190–191
cardiac function 52
cardiac output 4, 50, 51
cardiac troponin T (cTnT) 52
cardiovascular system
  responses to altitude 51t
Casas, M., et al. (2000) 212, 216t
catecholamines 61
cerebral edema 206
Chapman, R.F., et al. (1998) 11, 109, 111, 113t
Chapman, R.F., et al. (1999) 31, 34t
Chick, T.W., et al. (1993) 200, 201
Chung, D.-S., et al. (1995) 93t, 98t
citrate synthase (CS) 102, 169
Clarke, Ron xvii, xxi
cognitive function 67
colony forming unit-erythroid (CFU-E) 8
Colorado Altitude Room™ xxiv
Colorado Altitude Training (CAT) Hatch™ 206
Colorado Mountain Room™ 206, 218t
competition
  event duration and acclimatization 127f
  preparation by German athletes 157, 158f
  return to sea level 165
  U.S. preparation for 2002 Olympics 127, 130, 132–134
cortisol 57–58, 85
creatine phosphate 14
cross country skiers
  hypoxic apartment study 185–187
  oxidative stress 60
  skeletal muscle buffering 14
  training in nitrogen apartments 185
  $\dot{V}O_2$max study 25
cycling
  altitude and hour record 41f
  Australian study of hypoxic apartments 189
  Columbian athletes 171
  hypoxic apartment study 186–187, 191
  and IHE at rest 209
  and IHE during exercise 210–211
  LHTL and supplemental oxygen 200–201
  optimal altitude for time trials 39
  sample supplemental oxygen workout 201–203
  skeletal muscle buffering 14
  supplemental oxygen study 200–201
  supplemental oxygen workout 201–202
  time trials 39–41

D

Daniels, J. (1979) 124
Daniels, J., and N. Oldridge (1970) 25, 35t, 44t, 86, 90t, 120, 125t
Dehnert, C., et al. (2002) 112t
dehydration 53
Dill, D.B., and W.C. Adams (1971) 28, 35t, 36, 45t, 91t, 120, 121, 125t
doping, definition 217

Doubell, Ralph xvii
duration of altitude effect 226
duration of training
  general guidelines 181t
  length of stay at altitude 225–226
  in nitrogen apartment 192

E

ejection fraction (EF) 52
Ekblom, B., and B. Berglund (1991) 11, 12
endurance performance
  and altitude acclimatization 123
  and altitude exposure of trained endurance athletes 44–45t
  following rhEPO treatment 12t
  and intermittent hypoxic exposure at rest 208
enzymes
  adenosine triphosphatase 63
  in skeletal muscle and LHTH 101–104
epinephrine 61–62
  levels of women at altitude 64
EPO. *See* erythropoietin
erythrocytes. *See* red blood cells
erythropoiesis 8
erythropoietin (EPO)
  and altitude training xxiii
  artificial methods of increasing 10
  effects of rhEPO administration 12t
  and hypoxic apartments 185–186
  and intermittent exposure to simulated altitude 189–190
  and intermittent hypoxic exposure 207
  and LHTH strategy 89
  and LHTL strategy 109, 114
  mechanism 9f
  release and kidney oxygenation 8
ethical considerations 217
Ethiopian athletes
  medalists at Mexico City xix
  Olympic success xxii
Evans blue dye technique 96, 190–191
event duration and acclimatization 127f

F

fatigue
  and blood lactate levels 54
  and buffering capacity 17
  and oxidative stress 60
  and plasma glutamine 59
Faulkner, J.A., et al. (1967) 35t, 44t, 94t, 99t, 106t, 119, 125t
Faulkner, J.A., et al. (1968) 35t, 44t, 94t, 125t
female-specific responses to altitude 64–65
ferritin 56, 83
  and LHTH results 96
Fick equation 4, 5
figure skating 179
fitness level. *See* training status
fluid balance 53
fractional shortening 52

fraction of inspired oxygen 6
free radical production 60-61
Frey, W.O., et al. (2000) 214t
Friedmann, B., et al. (1999) 98t
Fulco, C.S., et al. (1998) 83
Fulco, C.S., et al. (2000) 83

G
Gammoudi, Mohamed xix
Gamow bag 206
gastrointestinal tract infections 58-59
Geiser, J., et al. (2001) 133
gender. *See* female-specific responses to altitude
genetics 14-15, 114
    identification of EPO gene 15
    and Kenyan athlete success 169
ginkgo biloba 66
glucose levels 55-56, 57
glycogenolysis 102
glycolysis 102
Glyde-Julian, C.G., et al. (2003) 209, 213t
Gore, C.J., et al. (1996) 24, 26t, 34t
Gore, C.J., et al. (1997) 24, 26t, 33, 34t, 44t
Gore, C.J., et al. (1998) 85, 91t, 100t, 226
Gore, C.J., et al. (2001) 14, 196t
Green, H.J., et al. (2000a) 63
Green, H.J., et al. (2000b) 63

H
H^+ buffering capacity 14, 17, 54, 187
Hahn, A.G. (1991) 83
Hahn, A.G., et al. (1992) 95t, 100t, 215t
Hamalainen, I.T., et al. (2000) 193
heart rate
    definition 4
    and non-training issues 229t
    at rest and during exercise at altitude 50
Hellemans, J. (1999) 208, 213t
hematocrit
    definition 5
    in elite aerobic athletes 96-97
    following rhEPO treatment 11, 12t
    and intermittent hypoxic exposure 212
    and intermittent hypoxic exposure at rest 208
    and intermittent hypoxic exposure during exercise 209
    and plasma volume 96
hematological effects of altitude training 3-13
    LHTH strategy 89, 92, 96-97
    in nitrogen apartment 185-187
hemoglobin
    and altitude training xxiii
    and anaerobic performance with LHTH 104-105
    of Columbian professional cyclists 171
    in elite aerobic athletes 96-97
    and ferritin levels 84
    following rhEPO treatment 11, 12t
    and intermittent hypoxic exposure 212
    and intermittent hypoxic exposure at rest 208

    and intermittent hypoxic exposure during exercise 209, 211
    and iron metabolism 56
    and LHTH strategy 87, 89
    and LHTL strategy 107
    measurement techniques 96, 190-191
    molecular structure 5-6
    normal levels 5
    oxygen-carrying capacity 6
    and plasma volume 96
    response to living at altitude xv
Hendriksen, I.J.M., and T. Meeuwsen (2003) 210, 214t
HiHiLo strategy 107, 109
Human Genome Project 14
Hutler, M., et al. (1998) 112t
hydration 53, 63-64, 229t
hypobaric chamber 111
    portable 206
    and pre-acclimatization 133
    training applications 218t
    used for acclimatization 123
hypoxia
    dosages and results 114
    via nitrogen dilution 111
hypoxia-inducible factor 1α (HIF-1α) 15
hypoxic apartment. *See* nitrogen apartment
Hypoxicator/Go2Altitude™ 208, 218t
Hypoxico Altitude Tent™ 206, 218t, xxiv
hypoxic sleeping units 111, 206-207
hypoxic stimulus 84
hypoxic ventilatory response (HVR) 52-53

I
immune system
    and non-training issues 229t
    postexercise suppression 57-58
    response to altitude 58-59
    suppression and overtraining 85
individuality of responses 85
    and hematological results with LHTH 96
    and LHTL strategy 109, 114
    and $\dot{V}O_2$max after acclimatization 122f
    of $\dot{V}O_2$max and altitude 30-32
infection 85, 114
infectious illness 58-59
inflammatory cytokines 85
Ingjer, F., and K. Myhre (1992) 93t, 99t
injury 85, 114
intensity of training 42-43, 46
    and acclimatization at Dr. Joe Vigil camp 142
    effect on research results 84
    of Kenyan distance runners 173
    and LHTL strategy 107
    modifications 226-228
    post-acclimatization 143
    and U.S. athlete training for 2002 Olympics 132
intermittent hypoxic exposure (IHE)
    combined with intermittent hypoxic training 212

intermittent hypoxic exposure (IHE) *(continued)*
    cycling 209, 210–211
    devices and methods 218*t*
    at rest 208–209
    at rest and during exercise 212
    scientific literature 213–216*t*
    simulated high elevations 210
    swimming 210
    triathletes 210, 211
interval training 107, 227*f*
iron (Fe) 6
iron (Fe) metabolism 56, 57*f*
iron (Fe) status
    effect on research results 83
    and LHTH results 96
    and LHTL strategy 114
    and non-training issues 229*t*

J
Jensen, K., et al. (1993) 34*t*, 44*t*, 93*t*, 122, 125*t*

K
Karlsen, T., et al. (2002) 215*t*
Katayama, K., et al. (1999) 211, 215*t*
kayaking 60
Kayser, B. (1996) 54
Keino, Kipchoge 166, xix, xxi
Kenyan athletes
    biomechanics and physiological economy 171
    blood lactate 171
    HADH levels 169
    medalists at Mexico City xix
    Olympic success 166, xxii
    physiological factors contributing to success 169–171
    training factors contributing to success 171–175
    $\dot{V}O_2$max 169
kidneys 8, 15
killer cell function 57
Kinsman, T.A., et al. (2002) 66
Klausen, T., et al. (1991) 98*t*
Kuno, S., et al. (1994) 102–103

L
lactate
    alactate speed training 161
    of Kenyan athletes 171
    and supplemental oxygen 193
    and training using hypoxic apartments 187
lactate dehydrogenase (LDH) 102
lactate paradox 53–54
Laitinen, H., et al. (1995) 185, 194*t*
Lawler, J., et al. (1988) 23, 27*t*
Levine, B. (2002) 83
Levine, B.D., and J. Stray-Gundersen (1997) 43, 94*t*, 99*t*, 106*t*, 112*t*
LHTH strategy
    and anaerobic performance 104–105, 106*t*
    and hematological factors 89, 92, 96–97, 98–100*t*
    and sea level endurance performance 86–89
    skeletal muscle factors 97, 101–104
LHTL strategy xxiii
    at Chula Vista, California 176
    diagram of model 108*f*
    differences in responses 11–12
    and intensity of training 107
    literature not in support 93–95*t*
    literature supporting 90–91*t*
    and normobaric hypoxia 183
    scientific studies 112–113*t*
    and supplemental oxygen 198–199
    using altitude tents 176
    using nitrogen-diluted apartments 177–178
    using nitrogen-diluted apartments (Australia) 189–192
    using nitrogen-diluted apartments (Finland) 185–187
    using nitrogen-diluted apartments (summary) 194–196*t*
    using nitrogen-diluted apartments (Sweden) 187–188
    using supplemental oxygen 176–177, 199–200
lipid utilization 55
Liu, Y., et al. (1998) 112*t*
"live high-train low" strategy. *See* LHTL strategy
live low-train high 208
locations
    for altitude training in U.S. 159
    for international altitude training 228, 230*t*
    for Kenyan distance runner training 172*t*

M
Madsen, Ørjan, PhD 152–155
Mann, Ron 162, 164–166
marathon performances 38*t*
marathon training 143
    McGee training program 165–166
Martin, D.T., et al. (2002) 195*t*
mathematical model
    of altitude effect on aerobic performance 37–38
    of altitude effect on anaerobic performance 48
    optimal elevation for cycling trials 39
Mattila, V., and H. Rusko (1996) 186, 196*t*
McClellan, T.M., et al. (1990) 49*t*
McClellan, T.M., et al. (1993) 49*t*
McGee, Bobby 162, 164–166
Meeuwsen, T., at al. (2001) 214*t*
menstrual cycle 65
metabolism
    carbohydrate utilization 55–56
    of iron 56, 57*f*
    and LHTH strategy 102
    substrate utilization 55–56
Mexico City Olympics 11–12, xvi–xx
mitochondria 13
    oxidation and LHTH strategy 102
    and pre-acclimatization 133
Mizuno, M., et al. (1990) 90*t*, 101, 103, 105, 106*t*
Morris, D.M., et al. (2000) 200, 201

Index

mountaineers
 and intermittent hypoxic exposure 212
 pre-acclimatization 133
muscle. *See* skeletal muscle

N

NADH (nicotinamide adenine dinucleotide) 193
natives of high altitude. *See also* Kenyan athletes
 cortisol levels 58
 domination in track and field events xxii
 responses to altitude 15
Niess, A.M., et al. (2003) 43
nitrogen apartment
 about 218t
 Australian studies 189–192
 description 183–184
 Finnish investigations 185–187
 and IHE during exercise 210–211
 increase in altitude studied 190–191
 length of training 192
 and non-hematological parameters 191
 summary of results 194–196t
 Swedish investigations 187–188
nitrogen house xxiv
non-training guidelines 132–133, 228, 229t
Nordic combined skiers 179
norepinephrine 61–62
 levels of women at altitude 64
normobaric hypoxia 183
normobaric hypoxic apartment. *See* nitrogen apartment
Norwegian altitude training program 152–155
 prior to Salt Lake City Olympics 177–178
Nummela, A., and H. Rusko (2000) 187, 194t
nutrition 63–64

O

Olympic performance
 Beamon, Bob xvi
 cycling 39
 Kenyan athletes 167–168t
 men's 10,000 m in Sydney xvi
 Mexico City xvi–xx
 records at Mexico City xvii
 results for Mexico City xviii–xix
 speedskating 40–41
 speedskating at Salt Lake City 42t
 swimmers after altitude training 160
 U.S. preparation for 2002 Olympics 127, 130, 132–134, 178–180
Olympic training centers
 Norway preparation for Salt Lake City 177–178
 U.S. at Chula Vista, California 176
 U.S. at Colorado Springs 176–177
 use of supplemental oxygen 197–198
optimal altitude 225
oral contraceptives 65
overtraining 85, 205
 avoiding with supplemental oxygen 205

avoiding with training modifications 226–228
 and LHTL strategy 114
oxidative stress 60–61, 229t
oxygen. *See also* supplemental oxygen
 concentration within nitrogen apartment 184
 delivery 4
 extraction 4
oxygen-carrying capacity xv
oxygen consumption. *See also* $\dot{V}O_2$max
 and altitude training 9
 definition 3
 factors affecting 4f
oxygen transport 6–7
oxyhemoglobin saturation. *See* S_AO_2

P

partial pressure of carbon dioxide 54
partial pressure of oxygen 6
 at different altitudes 8, 22t
 effect of altitude exposure 36
 and exposure to altitude 21
 and modifications to training 226
 and supplemental oxygen 198
performance. *See* athletic performance
Peronnet, F., et al. (1991) 29
pH 22f, 187
phosphofructokinase (PFK) 102
phosphorylase 102
Piehl-Aulin, K., et al. (1998) 188, 194t, 195t
Plaatjes, Mark 149–151
plasma volume 96
power output
 of cyclists at moderate altitude 47
 of cyclists living in hypoxic apartments 191
 of cyclists using supplemental oxygen 202
 of elite female cyclists 33
 following IHE regimen 208
 and IHE during exercise 209, 210, 211
 and supplemental oxygen 193, 205
pre-acclimatization 133–134
pressure. *See* partial pressure of oxygen
pressure gradients, and oxygen transport 6–7
Pugh, L.G.C.E. (1967) 120
pulmonary edema 206
pulmonary ventilation 52–53

R

Rahkila, P., and H. Rusko (1982) 94t, 106t
recombinant human EPO 217
recovery
 and non-training issues 229t
 and pre-acclimatization 134
 recommended at Barrios training camp 147
 and Skinner's altitude training for swimmers 161
 and U.S. athlete training for 2002 Olympics 132
red blood cell mass
 and anaerobic performance with LHTH 104–105

red blood cell mass *(continued)*
  and LHTH strategy 89, 92
  and LHTL strategy 107
  and training using hypoxic apartments 185–186
red blood cells
  and altitude training xxiii
  and erythropoietin 8
  and ferritin levels 84
  and hematocrit 5
  and hypoxic apartments 184
  and intermittent hypoxic exposure 207, 212
  maturation process 9
  production and overtraining 85
  profile of Columbian cyclists 171
  response to living at altitude xv
  sickle cell trait 66–67
Reeves, J.T., et al. (1992) 53
Reiss, Manfred, PhD 155–159
resources 228
respiratory water loss 53
responders/nonresponders 43
  hematological responses 11–12, 110f
  and LHTL strategy 109–110
reticulocytes
  and acute altitude exposure 89
  and intermittent hypoxic exposure 212
  and intermittent hypoxic exposure at rest 208
  sensitivity to erythropoiesis 189
  and simulated altitude exposure 185
return to sea level
  at Arturo Barrios training camp 147
  Dr. Joe Vigil recommendations 144
  general guidelines 181t
  Mark Plaatjes' program 150
  Norwegian altitude training program 154–155
  timing before competition 165
rhEPO. *See* erythropoietin
Roberges, R.A., et al. (1998) 32
Roberts, A.D., et al. (1998) 24, 26t, 34t, 44t
Roberts, A.D., et al. (2003) 191, 195t
Rodriguez, F.A., et al. (1999) 212, 213t, 216t
Rodriguez, F.A., et al. (2000) 208, 213t
Roi, G.S., et al. (1999) 45t
rowers 193
  and IHE during exercise 210–211
  oxidative stress 60
runners
  absolute training intensity 46t
  aerobic performance 33
  early research at altitude 119–120
  and intermittent exposure to simulated altitude 189–190
  and intermittent hypoxic exposure at rest 209
  LHTH and sea level performance 86–87
  and LHTH strategy 87
  and LHTL strategy 107
  postaltitude body composition 64
  serum cortisol study 58
  success of Kenyan athletes 166–169
  time loss vs. altitude 124, 128–129t
  training distance and time reduction 143t
  training intensity 43
  and training using hypoxic apartments 185–186
  $\dot{V}O_2$max study 25
running training strategies
  Arturo Barrios 144–149
  Dr. Joe Vigil 140–144
  Ronn Mann, Bobby McGee, Mark Wetmore 162, 164–166
running velocity 39, 43
  beneficial effects of LHTL 107
  and post-acclimatization training 143
  recommended at Barrios training camp 147
  sprints 47
  strategy for maintaining 164
Rusko, H.K., et al. (1995) 185, 194t
Rusko, H.K., et al. (1996) 58, 106t
Rusko, H.K., et al. (1999) 186, 194t

S

Saltin, B., C.K. Kim, et al. (1995) 97, 103, 169, 170
Saltin, B., H. Larsen, et al. (1995) 34t
Sandoval, D.A., and K.S. Matt (2003) 65
S_AO_2 21, 22t, 23
  and acute altitude exposure 26–27t
  and aerobic performance 33
  and decrease of $\dot{V}O_2$max 36f
Schmidt, W., et al. (2002) 171
sea level performance
  anaerobic 104–105
  hematological factors 89, 92, 96–97, 98–100t
  and LHTH strategy 86–89
  and LHTL strategy 105–111, 112–113t
  literature not in support of LHTH training 93–95t
  literature supporting LHTH training 90–91t
  skeletal muscle factors 97, 101–104
  $\dot{V}O_2$max study 92f
sickle cell trait 66–67
simulation of altitude change 114
  altitude tents 176
  contemporary methods and devices 218t
  hypoxic sleeping units 206–207
  nitrogen apartment 183–184
  supplemental oxygen 193–194
sites for altitude training 159, 172t, 228, 230t
skeletal muscle
  adenosine triphosphatase 63
  buffering capacity 14, 17
  capillarity and LHTL strategy 97, 101
  catabolism and overtraining 85
  composition of Kenyan runners 169
  effects of altitude training 13–14
  enzymes and LHTH strategy 101–104
  fatigue and blood lactate levels 54
  fatigue and supplemental oxygen 205
  and training using hypoxic apartments 191
Skinner, Jonty 159–162, 163t
sleep disturbances 65–66, 229t

speedskating
  ice surface and world records 39–41
  iron (Fe) metabolism study 56
  results at Salt Lake City Olympics 42*t*
  U.S. preparation for 2002 Olympics 179
splenic infarct 67
sprinting
  and acute exposure to altitude 105
  hypoxic apartment study 187
  and supplemental oxygen 193
  and training using hypoxic apartments 187
sprint velocity 47
Squires, R.W., and E.R. Buskirk (1982) 23, 24, 26*t*, 27*t*, 34*t*
Stray-Gundersen, J., et al. (1992) 56, 83, 109
Stray-Gundersen, J., et al. (2001) 113*t*
stress hormone response 57–58
stroke volume 4, 50–51
succinate dehydrogenase (SDH) 102
supplemental oxygen
  about 218*t*
  and cycling 201–203
  and LHTL strategy 176–177
  long-term training effect 200–201
  physiological effects 199–200
  scientific rationale for use 198*f*
  training considerations 205–206
  and triathletes 202–205
  use as training technique xxiv
Svedenhag, J., et al. (1991) 94*t*, 99*t*, 106*t*
Svedenhag, J., et al. (1997) 99*t*
swimming
  athletic performance at altitude 119
  and intermittent hypoxic exposure at rest 209
  and intermittent hypoxic exposure during exercise 210
  iron (Fe) metabolism study 56
  training program of Jonty Skinner 159–162, 163*t*
sympathetic nervous system 61–62

T

T-cell function 57
Telford, R.D., et al. (1996) 93*t*, 98*t*
Terrados, N., et al. (1988) 90*t*, 99*t*, 101, 103–104, 209, 214*t*
Townsend, N.E., et al. (2002) 192, 195*t*
training capacity 42–43, 46
training locations. *See* locations
training programs
  Arturo Barrios 144–149
  Dr. Joe Vigil 140–144, 145*t*
  guidelines and recommendations 181*t*
  Jonty Skinner 159–162, 163*t*
  of Kenyan distance runners 173, 174*t*
  Manfred Reiss, PhD 155–159
  Mark Plaatjes 149–151
  Norwegian altitude training program 152–155
training status
  effect on research results 84

and $\dot{V}O_2$max variability 30
triathletes
  hypoxic apartment study 186
  and IHE during exercise 210, 211
  skeletal muscle buffering 14
  and supplemental oxygen 202–205
Truijens, M.J., et al. (2003) 210, 215*t*

U

Uchakin, P., et al. (1995) 58
ultraviolet radiation 229*t*
upper respiratory infection 57, 58
urinary water loss 53
U.S. sites for altitude training 159

V

Vallier, J.M., et al. (1996) 35*t*, 36, 45*t*, 95*t*, 100*t*, 122, 126*t*, 211, 215*t*
Van Dyken, Amy 160
Vasankari, T.J., et al. (1993) 85
ventilatory responses 52–53
  of women 64
Vigil, Dr. Joe 140–144, 145*t*
$\dot{V}O_2$. *See* oxygen consumption
$\dot{V}O_2$max
  and acclimatization 125–126*t*
  vs. altitude elevation 28*f*
  and altitude exposure of trained endurance athletes 34–35*t*
  and altitude training xxiv
  definition 3
  following rhEPO treatment 10–11, 12*t*
  and intermittent hypoxic exposure 207
  and intermittent hypoxic exposure during exercise 210, 211
  in Kenyan distance runners 170
  and LHTH strategy 87, 88–89, 92*f*
  and LHTL strategy 109
  mathematical model of altitude effect 37
  and running at altitude 120, 121*f*
  and training using hypoxic apartments 186
volume of training 226–228

W

water loss 53
Web sites 228
Wetmore, Mark 162, 164–166
Weyand, P.G., et al. (1999) 49*t*
Wilber, R.L. (2001) 83
Wilber, R.L., et al. (2000) 57, 58
Wilber, R.L., et al. (2003b) 198
Wolski, L.A., et al. (1996) 83
world records
  cycling 39
  and LHTH strategy 87
  in Mexico City xvii
  speedskating 41
  sprinting 48

ABOUT THE AUTHOR

Randall L. Wilber, PhD, is a senior sport physiologist at the U.S. Olympic Training Center in Colorado Springs, Colorado (elevation 1,860 m/6,100 ft), where he oversees the operation of the Athlete Performance Laboratory. He has worked with U.S. Olympic athletes from a variety of sports and advised them on the scientific and practical aspects of altitude training. Those athletes include Lance Armstrong (five-time winner of the Tour de France, two-time Olympian), Alison Dunlap (2001 World Champion in cross-country mountain biking, two-time Olympian), Mari Holden (2000 World Champion in road cycling, 2000 Olympic silver medalist), Barb Lindquist (world-ranked #1 female triathlete in 2002 and 2003), Hunter Kemper (2003 Pan American Games gold medalist in triathlon, 2000 Olympian), Michael Phelps (four-time World Champion and four-time world-record holder at the 2003 Swimming World Championships), Apolo Ohno (2002 Olympic gold medalist in short track speedskating), Derek Parra (2002 Olympic gold medalist and world-record holder in long track speedskating), Christine Witty (2002 Olympic gold medalist and world-record holder in long track speedskating, three-time Olympian), and Johnny Spillane (2003 World Champion in Nordic combined skiing, 2002 Olympian).

Dr. Wilber's research interests include evaluating the effects of altitude training on athletic performance, exercise-induced asthma (EIA) in elite athletes, and the use of ergogenic aids for the enhancement of athletic performance. He has authored scientific papers on these topics that have been published in *Medicine and Science in Sports and Exercise, European Journal of Applied Physiology, Sports Medicine, International Journal of Sport Nutrition,* and *Journal of Strength and Conditioning Research.* Dr. Wilber has co-edited *Exercise-Induced Asthma: Pathophysiology and Treatment,* published by Human Kinetics in 2002. In addition, he has been an invited speaker at scientific meetings in Brazil, China, and Finland. Dr. Wilber was recognized as a Fellow of the American College of Sports Medicine (ACSM) in 1998.

Dr. Wilber holds a Masters degree and PhD in exercise physiology from Florida State University, where he conducted research on training and detraining in endurance athletes. Originally from the Pittsburgh, Pennsylvania area, Dr. Wilber spent most of his adult life in Florida prior to moving to Colorado in 1993. He has been involved in sports his entire life as an athlete (cross-country and track), coach, and exercise physiologist. Dr. Wilber's hobbies and recreational interests include running, biking, hiking, reading, and photography. He has completed the Leadville Trail 100 Mountain Bike Race (1999) and the Pikes Peak Ascent Running Race (2000-2002).